KU-454-908

Collins

GCSE Maths 2 tier-foundation for AQA B

HOMEWORK BOOK

BRIAN SPEED
KEITH GORDON
KEVIN EVANS

William Collins' dream of knowledge for all began with the publication of his first book in 1819. A self-educated mill worker, he not only enriched millions of lives, but also founded a flourishing publishing house. Today, staying true to this spirit, Collins books are packed with inspiration, innovation and practical expertise. They place you at the centre of a world of possibility and give you exactly what you need to explore it.

Collins. Do more.

Published by Collins
An imprint of HarperCollins*Publishers*
77–85 Fulham Palace Road
Hammersmith
London
W6 8JB

Browse the complete Collins catalogue at
www.collinseducation.com

© HarperCollins*Publishers* Limited 2006

10 9 8 7 6 5 4

ISBN-13 978-0-00-721571-3
ISBN-10 0-00-721571-1

Any educational institution that has purchased one copy of this publication may make unlimited duplicate copies for use exclusively within that institution. Permission does not extend to reproduction, storage within a retrieval system, or transmittal in any form or by any means, electronic, mechanical, photocopying, recording or otherwise, of duplicate copies for loaning, renting or selling to any other institution without the permission of the Publisher.

British Library Cataloguing in Publication Data
A Catalogue record for this publication is available from the British Library

Acknowledgements

Publishing Manager Michael Cotter

Cover design by Andy Parker and JPD

Internal design by JPD

Page make-up by JPD

Proofread by Marie Taylor

Production by Natasha Buckland

Printed and bound by Printing Express,
Hong Kong

Contents

Welcome to Collins GCSE Maths!

This homework book is designed to give you extra practice on the topics covered in class. Each exercise in the Collins GCSE Maths textbooks has a corresponding homework exercise of the same format in this book.

Occasionally, the homework may start with a worked example to ease you into the exercise and remind you of what you were doing in class. Included with this homework book is a CD-ROM containing versions of the Collins GCSE Maths textbook in electronic form. You may find it helpful to look at the relevant textbook page again, when doing your homework.

A few homework exercises include an "examination-type question" and they are highlighted with an asterisk * in the margin. These questions give you the opportunity to see how each topic might appear in your examinations.

We do hope you enjoy using Collins GCSE Maths, and we wish you every success in your studies!

Brian Speed, Keith Gordon, Kevin Evans

Developed by the team behind the highly popular and successful Maths Frameworking series, Collins GCSE Maths is the easiest way to achieve success in 2 tier Mathematics.

Instructions for using the CD-ROM

1. Insert the CD-ROM into your CD drive
2. The CD should run immediately
3. From the main menu, choose the textbook that you are using at school
4. Locate the Chapter you are working on, and open it.
5. Using the scroll bars to find the appropriate page you need.

If you experience any technical difficulties, contact **help@collinseducation.com** with details.

1 Statistical representation

HOMEWORK 1A

1 For the following surveys, decide whether the data should be collected by:
i sampling **ii** observation **iii** experiment.

a The number of 'doubles' obtained when throwing two dice.
b The number of people who use a zebra-crossing on a busy main road.
c People's choice of favourite restaurant.
d The makes of cars parked in the staff car-park.
e The number of times a 'head' appears when throwing a coin.
f The type of food students prefer to eat in the school canteen.

2 In a game, a fair six-sided dice has its faces numbered 0, 1 or 2.
The dice is thrown 36 times and the results are shown below.

2 0 2 2 1 2 0 2 2 0 0 2
1 2 2 2 0 2 2 0 1 2 2 1
0 2 2 0 2 0 2 2 0 1 2 0

a Copy and complete the frequency table for the data.

Number	Tally	Frequency
0		
1		
2		

b Based on the results in the table, how many times do you think each number appears on the dice?

★**3** The table shows the average highest daily temperature recorded during August in 24 cities around the world.

City	Temperature (°C)	City	Temperature (°C)
Athens	33	Madras	35
Auckland	15	Marrakesh	38
Bangkok	32	Moscow	22
Budapest	27	Narvik	16
Buenos Aires	16	Nice	29
Cape Town	18	Oporto	25
Dubai	39	Perth	17
Geneva	25	Pisa	30
Istanbul	28	Quebec	23
La Paz	17	Reykjavik	14
London	20	Tokyo	30
Luxor	41	Tripoli	31

a Copy and complete the grouped frequency table for the data.

Temperature (°C)	Tally	Frequency
11–15		
16–20		
21–25		
26–30		
31–35		
36–40		
41–45		

b In how many cities was the temperature higher than the temperature in London?

c Kay said that the difference between the highest and lowest temperatures was 34 °C but Derek said that it is was 27 °C. Explain how they obtained different answers.

★**4** Heather attends a Spanish evening class at her local college. One evening she conducted a survey of the ages of all the people who attended. She wrote down all the ages on a piece of paper as follows.

25 41 33 24 46
37 40 32 59
64 37 26 44
58 31 29 19
37 30 22
48 51 68 28 27
51 34 49

a How many people attended on that evening?

b Copy and complete the grouped frequency table for the data.

Age	Tally	Frequency
11–20		
21–30		
31–40		
41–50		
51–60		
61–70		

c How many people were aged under 21? Suggest a possible reason for this.

5 Pat measured the heights, to the nearest centimetre, of all the students in her class. Her data is given below.

143 135 147 153 146 138 151
142 139 131 144 127 143 145
140 143 153 141 150 137 136
125 136 140 131 147 154 142

a Draw a grouped frequency table for the data using class intervals 125–129, 130–134, 135–139, …

b In which interval do the most heights lie?

c How many students had a height of 140 cm or more?

HOMEWORK 1B

1 The pictogram shows the number of copies of *The Times* sold by a newsagent in a particular week.

		Total
Monday	▬ ▬ ▬	12
Tuesday	▬ ▬ ▬ ▬	
Wednesday	▬ ▬ ▪	
Thursday	▬ ▬ ▬ ▬	
Friday		
Saturday		

a How many newspapers does the symbol ▬ represent?

b Complete the totals for Tuesday, Wednesday and Thursday.

c The newsagent sold 15 copies on Friday and 22 copies on Saturday. Complete the pictogram for Friday and Saturday.

2 The pictogram shows the amount of sunshine in five English holiday resorts on one day in August.

Blackpool	Brighton	Scarborough	Skegness	Torbay
☼☼☼	☼◐	☼☼☼	☼☼	☼☼☼◐

Key: ☼ represents 3 hours.

a Write down the number of hours of sunshine for each resort.

b Great Yarmouth had $5\frac{1}{2}$ hours of sunshine on the same day. Explain why this would be difficult to show on this pictogram.

★3 The pictogram shows the number of call-outs five taxi drivers had on one evening.

Brian	✪ ✪
Mike	✪ ✪
Robert	✪ ◐
Steve	✪ ✪ ◐
Terry	✪ ◐

Key: ✪ represents 10 call-outs.

a How many call-outs did each taxi driver have?

b Explain why the symbol used in this pictogram is not really suitable.

c Joanne had 16 call-outs on the same evening. Redraw a suitable pictogram to show the call-outs for the six taxi drivers.

4 Rachel did a survey to show the number of people in each car on their way to work on a particular morning. This is a copy of her survey sheet.

No of people in each car	Frequency
1	30
2	19
3	12
4	5
5 or more	1

Draw a pictogram to illustrate her data.

HOMEWORK 1C

1 Linda asked a sample of people 'What is your favourite soap opera?'.
The bar chart shows their replies.

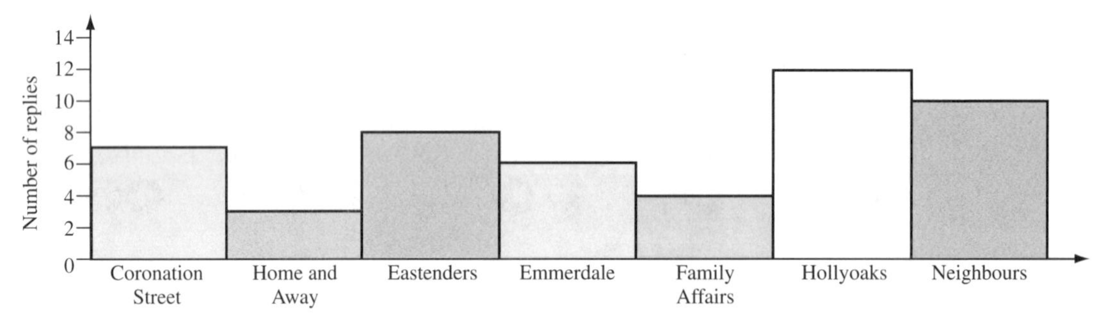

a Which soap opera got 6 replies?
b How many people were in Linda's sample?
c Linda collected the data from all her friends in Year 10 at school. Give two reasons why this is not a good way to collect the data.

2 The bar chart shows the results of a survey of shoe sizes in form 10KE.

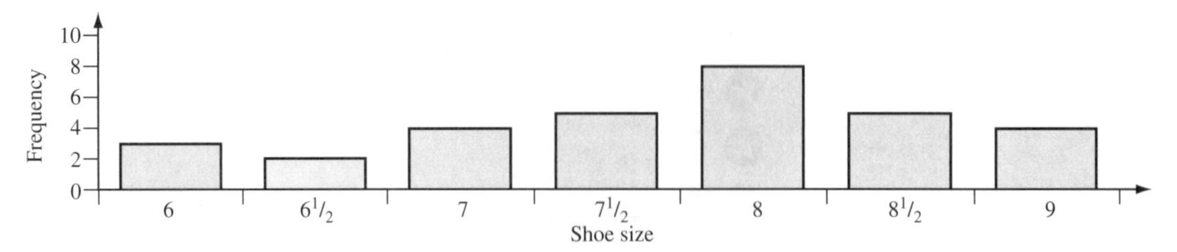

a How many students wear size $7\frac{1}{2}$ shoes?
b How many students were in the survey?
c What is the most common shoe size?
d Can you tell how many boys were in the survey? Explain your answer.

3 The table shows the lowest and highest marks six students got in a series of mental arithmetic tests.

	Abigail	Ben	Chris	Dave	Emma	Fay
Lowest mark	7	11	10	10	15	9
Highest mark	11	12	12	13	16	14

Draw a dual bar chart to illustrate the data.

4 The following data shows the times, to the nearest minute, that patients had to wait before seeing a doctor.

5	12	14	24	32	7	12	35	23	27	13	6
28	4	20	13	40	5	2	11	16	31	10	26
25	30	29	9	12	27	13	20	24	11	14	38

a Draw a grouped frequency table to show the waiting times of the patients, using class intervals 1–10, 11–20 , 21–30, 31–40.

b Draw a bar chart to illustrate the data.

5 Richard did a survey to find out which brand of crisps his friends preferred. He drew this bar chart to illustrate his data.

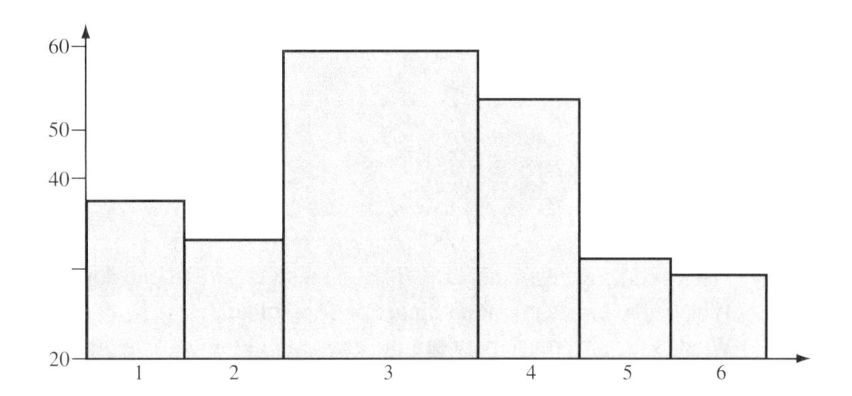

Richard's bar chart is very misleading. Explain how he could improve it and then redraw it, taking into account all your improvements.

HOMEWORK 1D

1 The line graph shows the monthly average exchange rate of the Japanese Yen for £1 over a six month period.

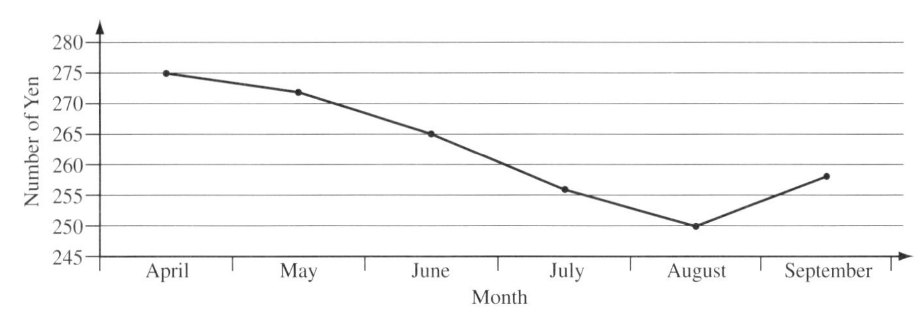

a In which month was the lowest exchange rate and what was that value?

b By how much did the exchange rate fall from April to August?

c Which month had the greatest fall in the exchange rate from the previous month?

d Mr Hargreaves changed £200 into yen during July. How many yen did he receive?

★**2** The table shows the temperature in Cuzco in Peru over a 24 hour period.

Time	**0000**	**0400**	**0800**	**1200**	**1600**	**2000**	**2400**
Temperature (°C)	1	–4	6	15	21	9	–1

a Draw a line graph for the data.

b From your graph estimate the temperature at 1800.

★3 The table shows the value, to the nearest million pounds, of a country's imports and exports.

Year	2001	2002	2003	2004	2005	2006
Imports	22	35	48	51	62	55
Exports	35	41	56	53	63	58

a Draw line graphs on the same axes to show the imports and exports of the country.
b Find the smallest and greatest difference between the imports and exports.

HOMEWORK 1E

1 The following stem-and-leaf diagram shows the number of TVs a retailer sold daily over a three week period.

1	2 8 9
2	0 2 4 4 4 4 5 7 8 8 9
3	1 2 4 8

Key 1 | 2 represents 14 TVs

a What is the greatest number of TVs the retailer sold in one day?
b What is the most common number of TVs sold daily?
c What is the difference between the greatest number and the least number of TVs sold?

2 The following stem-and-leaf diagram shows the ages of a group of people waiting for a train at a station.

1	6 8 9
2	4 7 8 9
3	0 2 4 5 6
4	2 5 5 6 8
5	0 4 8

Key 1 | 6 represents an age of 16

a How many people were waiting for a train?
b What is the age of the youngest person?
c What is the difference in age between the youngest person and oldest person?

3 A survey is carried out to find the speed, in miles per hour, of 30 vehicles travelling on a motorway. The results are shown below.

62 45 70 58 68 70 75 80 72 65 40 55 65 72 38
70 75 68 50 48 65 60 68 72 70 45 68 69 68 60

a Show the data on an ordered stem-and-leaf diagram. (Remember to show a key.)
b What is the most common speed?
c What is the difference between the greatest speed and the lowest speed?

2 Averages

HOMEWORK 2A

The mode is the value that occurs the most in a set of data. That is, it is the value with the highest frequency.

Example Terry scored the following number of goals in 12 school football matches:

1 2 1 0 1 0 0 1 2 1 0 2

The number which occurs most often in this list is 1. So, the mode is 1.
We can also say that the modal score is 1.

1 Find the mode for each set of data.

a 3, 1, 2, 5, 6, 4, 1, 5, 1, 3, 6, 1, 4, 2, 3, 2, 4, 2, 4, 2, 6, 5

b 17, 11, 12, 15, 11, 13, 18, 14, 17, 15, 13, 15, 16, 14

c 110, 10, 101, 10, 111, 110, 11, 101, 11, 111, 11, 101, 101, 111

d 1, –3, 3, 2, –1, 1, –3, –2, 3, –1, 2, 1, –1, 1, 2

e 7, $6\frac{1}{2}$, 6, $7\frac{1}{2}$, 8, $5\frac{1}{2}$, $6\frac{1}{2}$, 6, 7, $6\frac{1}{2}$, 7, $6\frac{1}{2}$, 6, $7\frac{1}{2}$

2 Find the modal category for each set of data.

a I, A, E, U, A, O, A, E, U, A, I, A, E, I, E, O, E, I, E, O

b ITV, C4, BBC1, C5, BBC2, C4, BBC1, C5, ITV, C4, BBC1, C4, ITV

c ↑, →, ↑, ←, ↓, →, ←, ↑, ←, →, ↓, ←, ←, ↑, →, ↓

d ♥, ♣, ♦, ♣, ♠, ♥, ♣, ♦, ♣, ♦, ♥, ♠

e ¥, €, £, €, $, £, ¥, €, £, $, €, £, $, €

3 Farmer Giles kept a record of the number of eggs his hens laid. His data is shown on the diagram below.

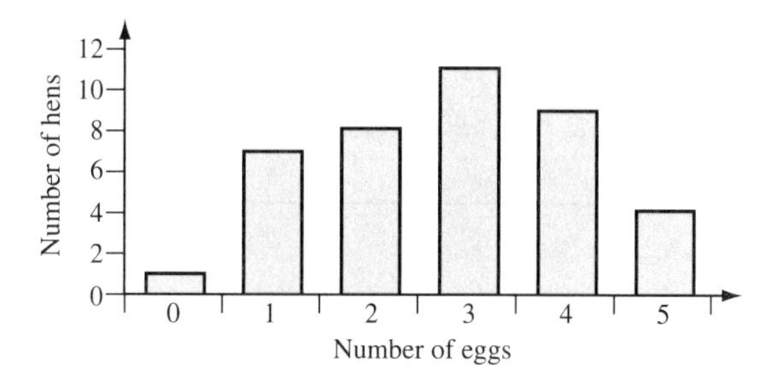

a How many hens did Farmer Giles have?

b What is the mode of the number of eggs laid?

c How many eggs were laid altogether?

★4 The grouped frequency table shows the number of marks a class of students obtained in a spelling test out of 30 marks.

a How many pupils are in the class?

b Write down the modal class for the number of marks.

c Stan looked at the table and said that at least one person got full marks. Explain why he may be wrong.

Number of marks	Frequency
1–5	1
6–10	2
11–15	4
16–20	8
21–25	10
26–30	5

★5 The data shows the times, to the nearest minute, that 30 shoppers had to wait in the queue at a checkout of a supermarket.

1	3	8	12	7	4	0	9	10	15	8	1	2	7	4
2	4	7	1	0	5	4	8	4	10	7	5	4	1	5

a Copy and complete the grouped frequency table.

Time in minutes	Tally	Frequency
0–3		
4–7		
8–11		
12–15		

b Draw a bar chart to illustrate the data.

c How many shoppers had to wait more than seven minutes?

d Write down the modal class for the time that the shoppers had to wait?

e How could the supermarket manager decrease the waiting time of the shoppers?

HOMEWORK 2B

The median is the value at the middle of a list of values after they have been put in order of size, from lowest to highest.

Example 1 Find the median for the list of numbers below.

2, 3, 5, 6, 1, 2, 3, 4, 5, 4, 6

Putting the list in numerical order gives

1, 2, 2, 3, 3, 4, 4, 5, 5, 6, 6

There are 11 numbers in the list, so the middle of the list is the 6th number.

Therefore, the median is 4.

Example 2 The ages of 20 people attending a conference are shown below.

28, 32, 46, 23, 28, 34, 52, 61, 45, 34, 39, 50, 26, 44, 60, 53, 31, 25, 37, 48

Draw a stem-and-leaf diagram and hence find the median age of the group.

Taking the tens to be the 'stem' and the units to be the 'leaves', the stem-and-leaf diagram is shown below. (**2** | 3 means 23)

Stem	Leaves
2	3 5 6 8 8
3	1 2 4 4 7 9
4	4 5 6 8
5	0 2 3
6	0 1

There is an even number of values in this list, so the middle of the list is between the two central values, 37 and 39. Therefore, the median is the value which is exactly midway between 37 and 39. Hence, the median is 38.

1 Find the median for each set of data.

a 18, 12, 15, 19, 13, 16, 10, 14, 17, 20, 11

b 22, 28, 42, 37, 26, 51, 30, 34, 43

c 1, –3, 0, 2, –4, 3, –1, 2, 0, 1, –2

d 12, 4, 16, 12, 14, 8, 10, 4, 6, 14

e 1.7, 2.1, 1.1, 2.7, 1.3, 0.9, 1.5, 1.8, 2.3, 1.4

2 The weights of eleven men in a local rugby team are shown below.

81 kg, 85 kg, 82 kg, 71 kg, 62 kg, 63 kg, 62 kg, 64 kg, 70 kg, 87 kg, 74 kg

a Find the median of their weights.

b Find the mode of their weights.

c Which is the better average to use? Explain your answer.

3 The bar chart shows the scores obtained in 20 throws of a dice.

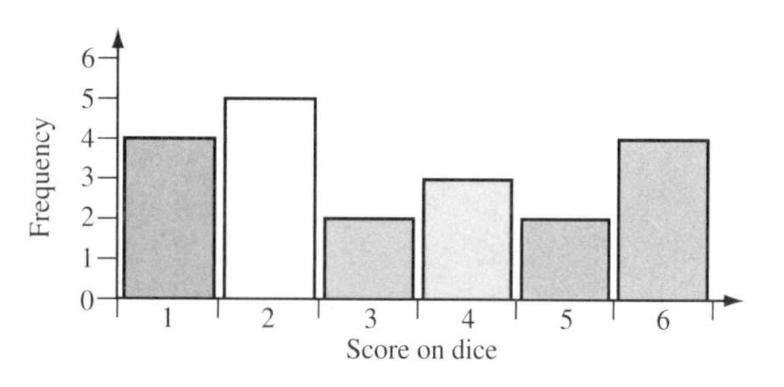

a Write down the modal score.

b Find the median score. Remember you must take into account all the scores.

c Do you think that the dice is biased? Explain your answer.

4 **a** Write down a list of seven numbers which has a median of 10 and a mode of 20.

b Write down a list of eight numbers which has a median of 10 and a mode of 20.

★**5** The marks of 21 students in a Science modular test are shown below.

45, 62, 27, 77, 40, 55, 80, 87, 49, 57, 35, 52, 59, 78, 48, 67, 43, 68, 38, 72, 81

Draw a stem-and-leaf diagram to find the median.

★**6** Jack is doing his Statistics Coursework based on 'Pulse Rates'. One of his hypotheses is that a person's pulse rate increases after exercise. To test his hypothesis he asks 31 students in a PE lesson to take their pulse rate at the start of the lesson and again at the end of the lesson. He records the data on an observation sheet and then illustrates it on a back to back stem-and-leaf diagram.

Before exercise		*After exercise*
9 9 8 5 3	**5**	
8 8 5 4 3 2 1	**6**	
9 8 7 7 4 2 2 2 0	**7**	1 1 2
9 6 5 2 2	**8**	0 0 2 7 8
8 7 5	**9**	0 2 2 3 5 5 8
4 2	**10**	1 2 4 5 8 9
	11	4 6 8 9
	12	2 5 6
	13	8 9
	14	4

(**7** | 2 means 72 beats per minute)

a Find the median pulse rate of the students at the start of the lesson.

b Find the median pulse rate of the students at the end of the lesson.

c What conclusions can Jack draw from the stem-and-leaf diagram?

HOMEWORK 2C

The mean of a set of data is the sum of all the values in the set divided by the total number of values in the set.

That is, mean $= \dfrac{\text{Sum of all values}}{\text{Total number of values}}$.

Example Find the mean of 4, 8, 7, 5, 9, 4, 8, 3.

Sum of all the values = 4 + 8 + 7 + 5 + 9 + 4 + 8 + 3 = 48
Total number of values = 8
Therefore, mean = 48 ÷ 8 = 6

1 Find the mean for each set of data.

a 4, 2, 5, 8, 6, 4, 2, 3, 5, 1
b 21, 25, 27, 20, 23, 26, 28, 22
c 324, 423, 342, 234, 432, 243
d 2.5, 3.6, 3.1, 4.2, 3.5, 2.9
e 1, 4, 3, 0, 1, 2, 5, 0, 2, 4, 2, 0

2 Calculate the mean for each set of data, giving your answer correct to one decimal place.

a 17, 24, 18, 32, 16, 28, 20
b 92, 101, 98, 102, 95, 104, 99, 96, 103
c 9.8, 9.3, 10.1, 8.7, 11.8, 10.5, 8.5
d 202, 212, 220, 102, 112, 201, 222
e 4, 2, –1, 0, 1, –3, 5, 0, –1, 4, –2, 1

3 A group of eight people took part in a marathon to raise money for charity. Their times to run the marathon were:

2 hours 40 minutes, 3 hours 6 minutes, 2 hours 50 minutes, 3 hours 25 minutes,
4 hours 32 minutes, 3 hours 47 minutes, 2 hours 46 minutes, 3 hours 18 minutes

Calculate their mean time in hours and minutes.

4 The monthly wages of 11 full-time staff who work in a restaurant are as follows:

£820, £520, £860, £2000, £800, £1600, £760, £810, £620, £570, £650

a Find their median wage.
b Calculate their mean wage.
c How many of the staff earn more than:
i the median wage **ii** the mean wage?
d Which is the better average to use? Give a reason for your answer.

★**5** The table shows the percentage marks which ten students obtained in Paper 1 and Paper 2 of their GCSE Mathematics examination.

	Ann	**Bridget**	**Carole**	**Daniel**	**Edwin**	**Fay**	**George**	**Hannah**	**Imman**	**Joseph**
Paper 1	72	61	43	92	56	62	73	56	38	67
Paper 2	81	57	49	85	62	61	70	66	48	51

a Calculate the mean mark for Paper 1.
b Calculate the mean mark for Paper 2.
c Which student obtained marks closest to the mean on both papers?
d How many students were above the mean mark on both papers?

★**6** The number of runs that a cricketer scored in seven innings were:

48, 32, 0, 62, 11, 21, 43

a Calculate the mean number of runs in the seven innings.
b After eight innings his mean score increased to 33 runs per innings. How many runs did he score in his eighth innings?

HOMEWORK 2D

The range for a set of data is the highest value in the set minus the lowest value in the set.

Example Rachel's marks in ten mental arithmetic tests were 4, 4, 7, 6, 6, 5, 7, 6, 9, 6.
Her mean mark is 60 ÷ 10 = 6 marks, and her range is 9 – 4 = 5 marks.
Robert's marks in the same tests were 6, 7, 6, 8, 5, 6, 5, 6, 5, 6.
His mean mark is 60 ÷ 10 = 6 marks, and his range is 8 – 5 = 3 marks.
Although the means are the same, Robert has a smaller range. This shows that Robert's results are more consistent.

1 Find the range for each set of data.

a 23, 18, 27, 14, 25, 19, 20, 26, 17, 24

b 92, 89, 101, 96, 100, 96, 102, 88, 99, 95

c 14, 30, 44, 25, 36, 27, 15, 42, 27, 12, 40, 31, 34, 24

d 3.2, 4.8, 5.7, 3.1, 3.8, 4.9, 5.8, 3.5, 5.6, 3.7

e 5, –4, 0, 2, –5, –1, 4, –3, 2, 2, 0, 1, –4, 0, –2

2 The table shows the ages of a group of students on an 'Outward Bound' course at a Youth Hostel.

Age	14	15	16	17	18	19
Number of students	2	3	8	5	6	1

a How many students were on the course?

b Write down the modal age of the students.

c What is the range of their ages?

d Draw a bar chart to illustrate the data.

3 A travel brochure shows the average monthly temperatures, in °C, for the island of Crete.

Month	April	May	June	July	August	September	October
Temperature °C	68	74	78	83	82	75	72

a Calculate the mean of these temperatures.

b Write down the range of these temperatures.

c The mean temperature for the island of Corfu was 77 °C and the range was 20 °C. Compare the temperatures for the two islands.

4 The table shows the daily attendance of three forms of 30 students over a week.

	Monday	**Tuesday**	**Wednesday**	**Thursday**	**Friday**
Form 10KG	25	25	26	27	27
Form 10RH	22	23	30	26	24
Form 10PB	24	29	28	25	29

a Calculate the mean attendance for each form.

b Write down the range for the attendance of each form.

c Which form had **i** the best attendance and **ii** the most consistent attendance? Give reasons for your answers.

★**5** The back to back stem-and-leaf diagram shows the marks of 30 pupils in one of their Key Stage 3 English tests.

Boys		*Girls*
	1	8
8 7 6 4	**2**	4 5 7 9
9 7 7 4 1 0	**3**	1 4 7 8
6 5 2 0	**4**	0 1 3 4 7
2	**5**	4

(**4** | 2 means 42 marks)

a Find the median mark for the boys and for the girls.
b Write down the range of the marks for the boys and for the girls.
c Compare the results of the boys and the girls.

HOMEWORK 2E

1 **a** For each set of data find the mode, the median and the mean.
i 6, 4, 5, 6, 2, 3, 2, 4, 5, 6, 1
ii 14, 15, 15, 16, 15, 15, 14, 16, 15, 16, 15
iii 31, 34, 33, 32, 46, 29, 30, 32, 31, 32, 33
b For each set of data decide which average is the best one to use and give a reason.

2 A supermarket sells oranges in bags of ten.
The weights of each orange in a selected bag are shown below.
134 g, 135 g, 142 g, 153 g, 156 g, 132 g, 135 g, 140 g, 148 g, 155 g
a Find the mode, the median and the mean for the weight of the oranges.
b The supermarket wanted to state the average weight on each bag they sold. Which of the three averages would you advise the supermarket to use? Explain why.

★**3** The weights, in kilograms, of a school football team are shown below.
68, 72, 74, 68, 71, 78, 53, 67, 72, 77, 70
a Find the median weight of the team.
b Find the mean weight of the team.
c Which average is the better one to use? Explain why.

★**4** Jez is a member of a pub quiz team and, in the last eight games, his total points are shown below.
62, 58, 24, 47, 64, 52, 60, 65
a Find the median for the number of points he scored over the eight games.
b Find the mean for the number of points he scored over the eight games.
c The team captain wanted to know the average for each member of the team. Which average would Jez use? Give a reason for your answer.

HOMEWORK 2F

1 Find **i** the mode, **ii** the median and **iii** the mean from each frequency table below.
a A survey of the collar sizes of all the male staff in a school gave these results.

Collar size	12	13	14	15	16	17	18
Number of staff	1	3	12	21	22	8	1

b A survey of the number of TVs in pupils' homes gave these results.

Number of TVs	1	2	3	4	5	6	7
Frequency	12	17	30	71	96	74	25

2 A survey of the number of pets in each family of a school gave these results.

Number of pets	0	1	2	3	4	5
Frequency	28	114	108	16	15	8

a Each child at the school is shown in the data, how many children are at the school?
b Calculate the median number of pets in a family.
c How many families have less than the median number of pets?
d Calculate the mean number of pets in a family. Give your answer to 1 dp.

HOMEWORK 2G

1 Find for each table of values given below **i** the modal group and **ii** an estimate for the mean.

a

Score	0 – 20	21 – 40	41 – 60	61 – 80	81 – 100
Frequency	9	13	21	34	17

b

Cost (£)	0.00 – 10.00	10.01 – 20.00	20.01 – 30.00	30.01 – 40.00	40.01 – 60.00
Frequency	9	17	27	21	14

2 A survey was made to see how long casualty patients had to wait before seeing a doctor. The following table summarises the results for one shift.

Time (minutes)	0 – 10	11 – 20	21 – 30	31 – 40	41 – 50	51 – 60	61 – 70
Frequency	1	12	24	15	13	9	5

a How many patients were seen by a doctor in the survey of this shift?
b Estimate the mean waiting time taken per patient.
c Which average would the hospital use for the average waiting time?

HOMEWORK 2H

1 The table shows the number of goals scored by a football team in 20 matches.

Goals	0	1	2	3	4
Frequency	5	7	4	3	1

a Draw a frequency polygon to illustrate the data.
b Calculate the mean number of goals scored per game.

2 The table shows the times taken by 50 pupils to complete a multiplication square.

Time, s, seconds	$10 < s \leqslant 20$	$20 < s \leqslant 30$	$30 < s \leqslant 40$	$40 < s \leqslant 50$	$50 < s \leqslant 60$
Frequency	4	10	16	12	8

a Draw a frequency polygon to illustrate the data.
b Calculate an estimate for the mean time taken by the pupils.

3 The waiting times for customers at a supermarket checkout are shown in the table.

Time, m, minutes	$0 < m \leqslant 2$	$2 < m \leqslant 4$	$4 < m \leqslant 6$	$6 < m \leqslant 8$	$8 < m \leqslant 10$
Frequency	3	5	10	8	4

a Draw a frequency polygon to illustrate the data.
b Calculate an estimate for the mean waiting time for the customers.

3 Probability

HOMEWORK 3A

1 When throwing a fair dice, state whether each of the following events are impossible, very unlikely, unlikely, evens, likely, very likely or certain.
a The score is a factor of 20.
b The score is $3\frac{1}{2}$.
c The score is a number less than six.
d The score is a one.
e The score is a number greater than zero.
f The score is an odd number.
g The score is a multiple of three.

2 Draw a probability scale and put an arrow to show approximately the probability of each of the following events happening.
a It will snow on Christmas Day this year.
b The sun will rise tomorrow morning.
c Someone in your class will have a birthday this month.
d It will rain tomorrow.
e Someone will win the Jackpot in the National Lottery this week.

3 Give an event of your own where you think the probability is:
a impossible **b** very unlikely **c** unlikely **d** evens
e likely **f** very likely **g** certain.

HOMEWORK 3B

Example A bag contains five red balls and three blue balls. A ball is taken out at random. What is the probability that it is: **a** red **b** blue **c** green?

a There are five red balls out of a total of eight, so P(red) = $\frac{5}{8}$.
b There are three blue balls out of a total of eight, so P(blue) = $\frac{3}{8}$.
c There are no green balls, so P(green) = 0.

1 When drawing a card from a well shuffled pack of cards, what is the probability of each of the following events? Remember to cancel down the probability fraction if possible.
a Drawing an Ace. **b** Drawing a picture card.
c Drawing a Diamond. **d** Drawing a Queen or a King.
e Drawing the Ace of Spades. **f** Drawing a red Jack.
g Drawing a Club or a Heart.

2 The numbers 1 to 10 inclusive are placed in a hat. Irene takes a number out of the bag without looking. What is the probability that she draws:

a the number 10 **b** an odd number **c** a number greater than 4
d a prime number **e** a number between 5 and 9?

3 A bag contains two blue balls, three red balls and four green balls. Frank takes a ball from the bag without looking. What is the probability that he takes out:

a a blue ball **b** a red ball **c** a ball that is not green **d** a yellow ball?

4 In a prize raffle there are 50 tickets: 10 coloured red, 10 coloured blue and the rest coloured white. What is the probability that the first ticket drawn out is:

a red **b** blue **c** white **d** red or white **e** not blue?

5 A bag contains 15 coloured balls. Three are red, five are blue and the rest are black. Paul takes a ball at random from the bag.

a Find:
i P (he chooses a red) **ii** P (he chooses a blue) **iii** P(he chooses a black).

b Add together the three probabilities. What do you notice?

c Explain your answer to part **b**.

6 Boris knows that when he plays a game of chess, he has a 65% chance of winning a game and a 15% chance of losing a game. What is the probability that he draws a game?

HOMEWORK 3C

Example What is the probability of not picking an Ace from a pack of cards?

First, find the probability of picking an Ace: P (picking an Ace) = $\frac{4}{52} = \frac{1}{13}$.
Therefore, P (not picking an Ace) = $1 - \frac{1}{13} = \frac{12}{13}$.

1 a The probability of winning a prize in a tombola is $\frac{1}{20}$. What is the probability of not winning a prize in the tombola?

b The probability that it will rain tomorrow is 65%. What is the probability that it will not rain tomorrow?

c The probability that Josie wins a game of tennis is 0.8. What is the probability that she loses a game?

d The probability of getting a double six when throwing two dice is $\frac{1}{36}$. What is the probability of not getting a double six?

2 Harvinder picks a card from a pack of well-shuffled playing cards. Find the probability that she picks:

a **i** a King **ii** a card that is not a King
b **i** a Spade **ii** a card that is not a Spade
c **i** a 9 or a 10 **ii** neither a 9 nor a 10.

★**3** The following letters are put into a bag.

a Stan takes a letter at random. What is the probability that:
i he takes a letter A **ii** he does not take a letter A?

b Pat takes an R and keeps it. Stan now takes a letter from those remaining.
i What is the probability that he takes a letter A?
ii What is the probability that he does not take a letter A?

HOMEWORK 3D

1 Shaheeb throws an ordinary dice. What is the probability that he throws:

a an even number **b** 5 **c** an even number or 5?

2 Jane draws a card from a pack of cards. What is the probability that she draws:

a a red card **b** a black card **c** a red or a black card?

3 Natalie draws a card from a pack of cards. What is the probability that she draws one of the following numbers?

a Ace **b** King **c** Ace or King

4 A letter is chosen at random from the letters in the word STATISTICS. What is the probability that the letter will be:

a S **b** a vowel **c** S or a vowel?

5 A bag contains 10 white balls, 12 black balls and 8 red balls. A ball is drawn at random from the bag. What is the probability that it will be:

a white **b** black **c** black or white

d not red **e** not red or black?

6 A spinner has numbers and colours on it, as shown in the diagram. Their probabilities are given in the tables.

Red	0.5
Green	0.25
Blue	0.25

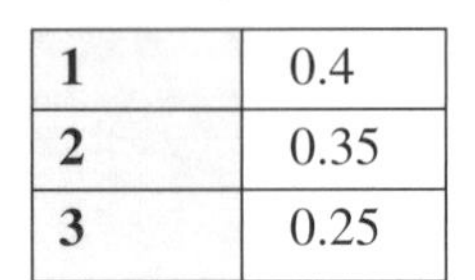

1	0.4
2	0.35
3	0.25

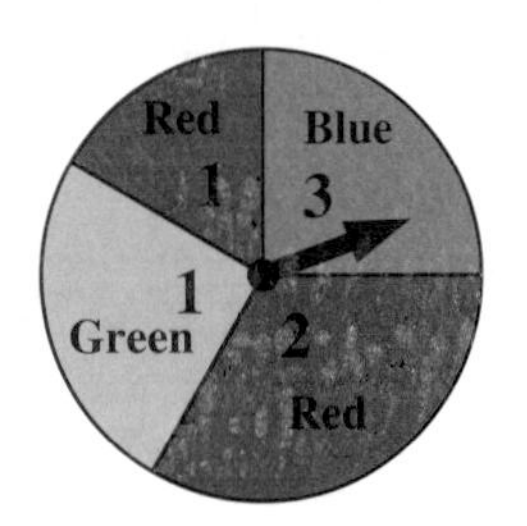

When the spinner is spun what is the probability of each of the following?

a Red or blue **b** 2 or 3 **c** 3 or blue **d** 2 or green

e **i** Explain why the answer to **c** is 0.25 and not 0.5.

ii What is the answer to P(2 or red)?

7 Debbie has ten CDs in her multi-changer, four of which are rock, two are dance and four are classical. She puts the player on random play. What is the probability that the first CD will be:

a rock or dance **b** rock or classical **c** not rock?

8 Frank buys a dozen free-range eggs. The farmer tells him that a quarter of the eggs his hens lay have double yolks.

a How many eggs with double yolks can Frank expect to get?

b He cooks three and finds they all have a single yolk. He argues that he now has a 1 in 3 chance of a double yolk from the remaining eggs. Explain why he is wrong.

★**9** John has a bag containing six red, five blue and four green balls. One ball is picked from the bag at random. What is the probability that the ball is:

a red or blue **b** not blue **c** pink **d** red or not blue?

HOMEWORK 3E

1 Katrina throws two dice and records the number of doubles that she gets after various numbers of throws. The table shows her results.

Number of throws	10	20	30	50	100	200	600
Number of doubles	2	3	6	9	17	35	102

a Calculate the experimental probability of a double at each stage that Katrina recorded her results.

b What do you think the theoretical probability is for the number of doubles when throwing two dice?

2 Mary made a six-sided spinner, like the one shown in the diagram. She used it to play a board game with her friend Jane. The girls thought that the spinner wasn't very fair as it seemed to land on some numbers more than others. They spun the spinner 120 times and recorded the results. The results are shown in the table.

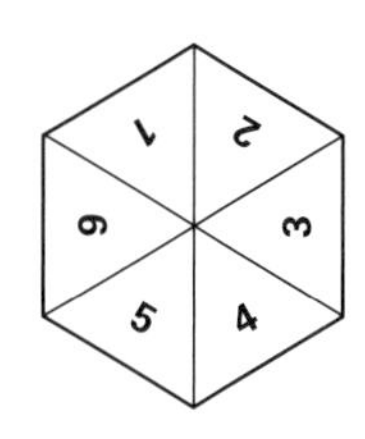

Number spinner lands on	1	2	3	4	5	6
Number of times	22	17	21	18	26	16

a Work out the experimental probability of each number.

b How many times would you expect each number to occur if the spinner is fair?

c Do you think that the spinner is fair? Give a reason for your answer.

3 In a game at the fairground a player rolls a coin onto a squared board with some of the squares coloured blue, green or red. If the coin lands completely within one of the coloured squares the player wins a prize. The table below shows the probabilities of the coin landing completely within a winning colour.

Colour	Blue	Green	Red
Probability	0.3	0.2	0.1

a On one afternoon 300 games were played. How many coins would you expect to land on: **i** a blue square **ii** a green square **iii** a red square?

b What is the probability that a player loses a game?

HOMEWORK 3F

1 Copy and complete the sample space diagram to show the total score when two dice are thrown together.

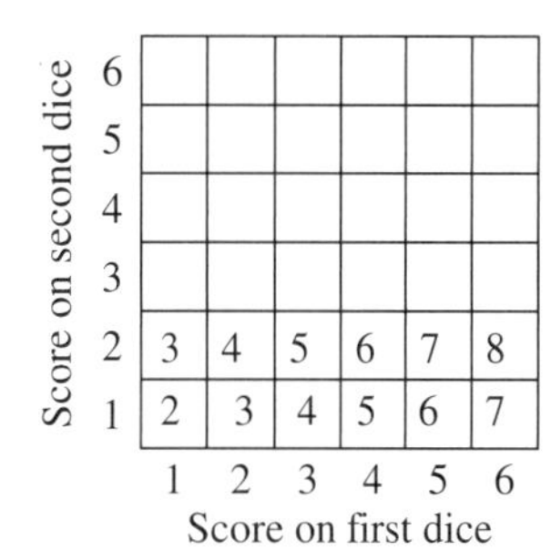

a What is the most likely score?

b Which two scores are least likely?

c Write down the probability of getting a double six.

d What is the probability that a score is:

i 11

ii 4

iii greater than 9

iv an odd number

v 4 or less

vi a multiple of 4?

2 Copy and complete the sample space diagram to show the outcomes when a dice and a coin are thrown together.

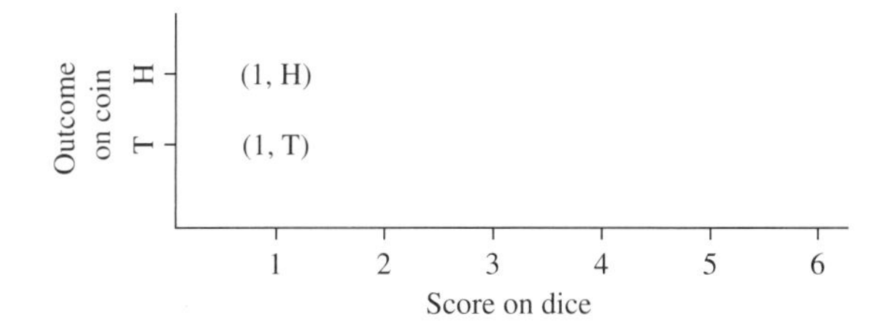

Find the probability of getting:

a a Head and a score of 6

b a Tail and an even score

c a score of 3.

★**3** Elaine throws a coin and spins a 5-sided spinner. One possible outcome is (Heads, 5).

a List all the possible outcomes.

b What is the probability of getting Tails on the coin and an odd number on the spinner?

★**4** A bag contains five discs that are numbered 2, 4, 6, 8 and 10. Sharleen takes a disc at random from the bag and puts the disc back. She shakes the bag and takes a disc again. She adds together the two numbers on the discs she has chosen.

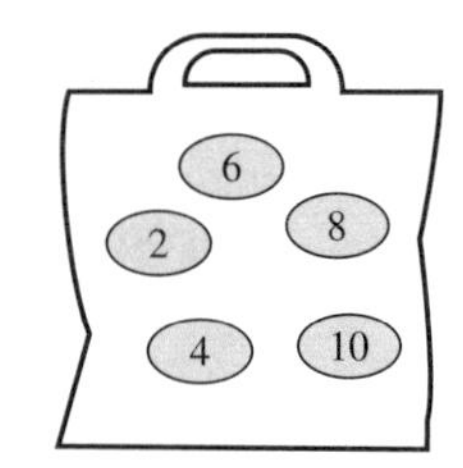

a Copy and complete the table to show all the possible totals.

First number

Second number

	2	**4**	**6**	**8**	**10**
2					
4					
6					
8					
10					

b Find the probability that the total is:

i 12 **ii** 20 **iii** 15 **iv** a square number **v** a multiple of 3.

HOMEWORK 3G

1 I throw an ordinary dice 600 times. How many times can I expect to get a score of 1?

2 I toss a coin 500 times. How many times can I expect to get a tail?

3 I draw a card from a pack of cards and replace it. I do this 104 times. How many times would I expect to get:

a a red card **b** a Queen **c** a red seven **d** the Jack of Diamonds?

4 The ball in a roulette wheel can land on any number from 0 to 36. I always bet on the same block of numbers 0–6. If I play all evening and there is a total of 111 spins of the wheel in that time, how many times could I expect to win?

5 I have five tickets for a raffle and I know that the probability of my winning the prize is 0.003. How many tickets were sold altogether in the raffle?

6 In a bag there are 20 balls, ten of which are red, three yellow, and seven blue. A ball is taken out at random and then replaced. This is repeated 200 times. How many times would I expect to get:

a a red ball **b** a yellow or blue ball

c a ball that is not blue **d** a green ball?

7 A sampling bottle contains black and white balls. It is known that the probability of getting a black ball is 0.4. How many white balls would you expect to get in 200 samples?

8 **a** Fred is about to take his driving test. The chance he passes is $\frac{1}{3}$. His sister says 'Don't worry if you fail because you are sure to pass within three attempts because $3 \times \frac{1}{3} = 1$'. Explain why his sister is wrong.

b If Fred does fail would you expect the chance that he passes next time to increase or decrease? Explain your answer.

★**9** An opinion poll used a sample of 200 voters in one area. 112 said they would vote for Party A. There are a total of 50 000 voters in the area.

a If they all voted, how many would you expect to vote for Party A?

b The poll is accurate within 10%. Can Party A be confident of winning?

HOMEWORK 3H

1 Two dice are thrown together. Draw a probability diagram to show the total score.

a What is the probability of a score that is:

i 7 **ii** 5 or 8 **iii** bigger than 9 **iv** between 2 and 5

v odd **vi** a non-square number?

2 Two dice are thrown. Draw a probability diagram to show the outcomes as a pair of co-ordinates.

What is the probability that:

a the score is a 'double'

b at least one of the dice shows 3

c the score on one die is three times the score on the other die

d at least one of the dice shows an odd number

e both dice show a 5

f at least one of the dice will show a 5

g exactly one die shows a 5?

3 Two dice are thrown. The score on the first die is doubled and the score on the second die is subtracted.

Complete the probability space diagram.

For the event described above, what is the probability of a score of:

a 1

b a negative number

c an even number

d 0 or 1

e a prime number?

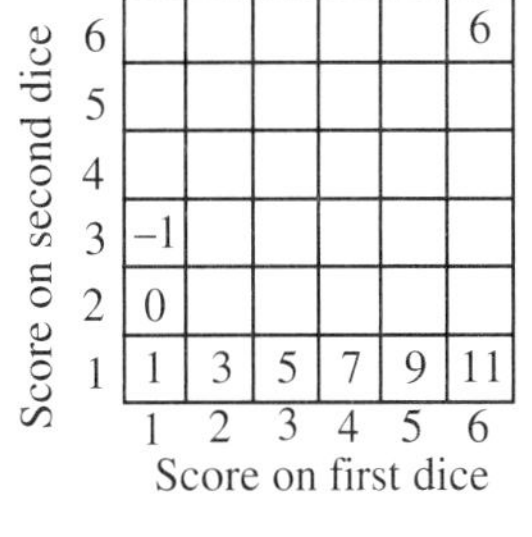

4 When two coins are tossed together, what is the probability of:

a 2 heads or 2 tails **b** a head and a tail **c** at least 1 head?

5 When three coins are tossed together, what is the probability of:

a 3 heads or 3 tails **b** 2 tails and 1 head **c** at least 1 head?

6 When a dice and a coin are thrown together, what is the probability of each of the following outcomes?

a You get a tail on the coin and a 3 on the dice.

b You get a head on the coin and an odd number on the dice.

★7 Max buys two bags of bulbs from his local garden centre. Each bag has 4 bulbs. Two bulbs are daffodils, one is a tulip and one is a hyacinth. Max takes one bulb from each bag.

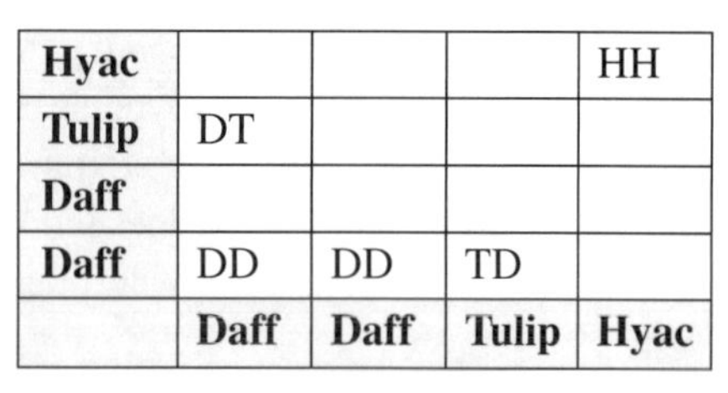

Hyac				HH
Tulip	DT			
Daff				
Daff	DD	DD	TD	
	Daff	**Daff**	**Tulip**	**Hyac**

a There are six possible different pairs of bulbs. List them all.

b Complete the sample space diagram.

c What is the probability of getting two daffodil bulbs?

d Explain why the answer is not $\frac{1}{6}$.

4 Pie charts, scatter diagrams and surveys

HOMEWORK 4A

1 The table shows the time taken by 60 people to travel to work.

Time in minutes	10 or less	Between 10 and 30	30 or more
Frequency	8	19	33

Draw a pie chart to illustrate the data.

2 The table shows the number of GCSE passes that 180 students obtained.

GCSE passes	9 or more	7 or 8	5 or 6	4 or less
Frequency	20	100	50	10

Draw a pie chart to illustrate the data.

3 Tom is doing a statistics project on the use of computers. He decides to do a survey to find out the main use of computers by 36 of his school friends. His results are shown in the table.

Main use	e-mail	Internet	Word processing	Games
Frequency	5	13	3	15

a Draw a pie chart to illustrate his data.

b What conclusions can you draw from his data?

c Give reasons why Tom's data is not really suitable for his project.

★4 In a survey, a TV researcher asks 120 people at a leisure centre to name their favourite type of television programme. The results are shown in the table.

Type of programme	Comedy	Drama	Films	Soaps	Sport
Frequency	18	11	21	26	44

a Draw a pie chart to illustrate the data.

b Do you think the sample chosen by the researcher is representative of the population? Give a reason for your answer.

★5 Marion is writing an article on health for a magazine. She asked a sample of people the question: 'When planning your diet, do you consider your health?' The pie chart shows the results of her question.

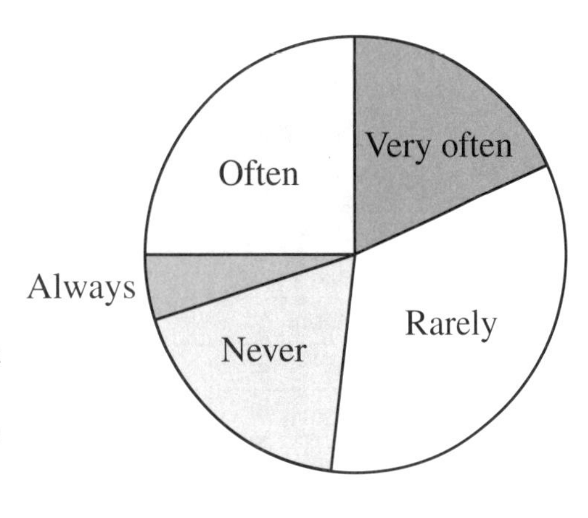

- **a** What percentage of the sample responded 'often'.
- **b** What response was given by about a third of the sample?
- **c** Can you tell how many people there were in the sample? Give a reason for your answer.
- **d** What other questions could Marion ask?

HOMEWORK 4B

★1 The table below shows the heights and weights of twelve students in a class.

Student	Weight (Kg)	Height (cm)
Ann	51	123
Bridie	58	125
Ciri	57.5	127
Di	62	128
Emma	59.5	129
Flo	65	129
Gill	65	133
Hanna	65.5	135
Ivy	71	137
Joy	75.5	140
Keri	70	143
Laura	78	145

- **a** Plot the data on a scatter diagram.
- **b** Draw the line of best fit.
- **c** Jayne was absent from the class, but they knew she was 132 cm tall. Use the line of best fit to estimate her weight.
- **d** A new girl joined the class who weighed 55 kg. What height would you expect her to be?

★**2** The table below shows the marks for ten pupils in their mathematics and music examinations.

Pupil	Maths	Music
Alex	52	50
Ben	42	52
Chris	65	60
Don	60	59
Ellie	77	61
Fan	83	74
Gary	78	64
Hazel	87	68
Irene	29	26
Jez	53	45

a Plot the data on a scatter diagram. Take the x-axis for the mathematics scores and mark it from 20 to 100. Take the y-axis for the music scores and mark it from 20 to 100.

b Draw the line of best fit.

c One of the pupils was ill when they took the music examination. Which pupil was it most likely to be?

d Another pupil, Kris, was absent for the music examination but scored 45 in mathematics, what mark would you expect him to have got in music?

e Another pupil, Lex, was absent for the mathematics examination but scored 78 in music, what mark would you expect him to have got in mathematics?

HOMEWORK 4C

1 'People like the video rental store to be open 24 hours a day.'

a To see whether this statement is true, design a data collection sheet which will allow you to capture data while standing outside a video rental store.

b Does it matter at which time you collect your data?

2 The youth club wanted to know which types of activities it should plan, e.g. craft, swimming, squash, walking, disco etc.

a Design a data collection sheet which you could use to ask the pupils in your school which activities they would want in a youth club.

b Invent the first 30 entries on the chart.

★**3** What types of film do your age group watch at the cinema the most? Is it comedy, romance, sci-fi, action, suspense or something else?

a Design a data collection sheet to be used in a survey of your age group.

b Invent the first thirty entries on your sheet.

4 Design a two-way table to show the type of music students prefer to listen to in different year groups.

HOMEWORK 4D

1 Design a questionnaire to test the following statement.
'Young people aged 16 and under will not tell their parents when they have been drinking alcohol, but the over 16s will always let their parents know.'

★**2** ‘Boys will use the Internet almost everyday but girls will only use it about once a week.’ Design a questionnaire to test this statement.

★**3** Design a questionnaire to test the following hypothesis.
‘When you are in your twenties, you watch less TV than any other age group.’

4 While on holiday in Wales, I noticed that in the supermarkets there were a lot more women than men, and even then, the only men I did see were over 65.

a Write down a hypothesis from the above observation.

b Design a questionnaire to test your hypothesis.

5 Basic number

HOMEWORK 5A

1 Find the row and column sums of each of these grids.

a

2	5	1	□
8	3	4	□
9	7	6	□
□	□	□	□

b

3	0	1	□
7	5	6	□
8	4	2	□
□	□	□	□

c

0	3	5	□
9	6	8	□
2	7	1	□
□	□	□	□

d

3	1	8	□
6	2	0	□
7	9	4	□
□	□	□	□

e

8	1	3	□
6	4	5	□
2	7	0	□
□	□	□	□

2 Find the numbers missing from each of these grids. Remember: the numbers missing from each grid must be chosen from 0 to 9 without any repeats.

a

1	5	□	10
8	□	7	17
□	9	6	18
12	16	17	45

b

4	□	5	10
□	6	□	8
7	8	□	18
13	15	8	36

c

□	□	□	14
0	9	8	17
2	□	5	10
9	18	14	□

d

4	□	□	12
□	0	5	12
□	8	3	□
12	□	10	36

e

□	□	3	10
□	0	□	□
8	2	□	15
16	□	12	□

HOMEWORK 5B

1 Write down the answer to each of the following without looking at a multiplication square.

a 4×3	**b** 7×4	**c** 6×5	**d** 3×8	**e** 8×6
f 9×4	**g** 5×9	**h** 9×7	**i** 8×8	**j** 9×8
k 7×6	**l** 7×7	**m** 4×6	**n** 8×7	**o** 5×5

2 Write down the answer to each of the following without looking at a multiplication square.

a $14 \div 2$	**b** $28 \div 4$	**c** $24 \div 6$	**d** $20 \div 5$	**e** $18 \div 3$
f $35 \div 5$	**g** $27 \div 3$	**h** $32 \div 4$	**i** $24 \div 8$	**j** $21 \div 7$
k $42 \div 6$	**l** $40 \div 8$	**m** $18 \div 9$	**n** $49 \div 7$	**o** $48 \div 6$

3 Write down the answer to each of the following. Look carefully at the signs, because they are a mixture of +, –, × and ÷.

a $8 + 5$ **b** $20 - 6$ **c** 4×5 **d** $16 \div 4$ **e** $14 - 8$

f $15 \div 3$ **g** $16 + 8$ **h** 5×7 **i** $16 + 5$ **j** $36 \div 6$

k $17 - 8$ **l** 9×3 **m** $42 \div 7$ **n** 6×9 **o** $21 - 6$

4 Write down the answer to each of the following.

a 4×10 **b** 7×10 **c** 9×10 **d** 11×10 **e** 3×100

f 5×100 **g** 24×100 **h** 45×100 **i** $80 \div 10$ **j** $130 \div 10$

k $510 \div 10$ **l** $1000 \div 10$ **m** $700 \div 100$ **n** $900 \div 100$ **o** $1200 \div 100$

HOMEWORK 5C

1 Work out each of these.

a $3 \times 4 + 7 =$ **b** $8 + 2 \times 4 =$ **c** $12 \div 3 + 4 =$ **d** $10 - 8 \div 2 =$

e $7 + 2 - 3 =$ **f** $5 \times 4 - 8 =$ **g** $9 + 10 \div 5 =$ **h** $11 - 9 \div 1 =$

i $12 \div 1 - 6 =$ **j** $4 + 4 \times 4 =$

2 Work out each of these. Remember: first work out the bracket.

a $3 \times (2 + 4) =$ **b** $12 \div (4 + 2) =$ **c** $(4 + 6) \div 5 =$

d $(10 - 6) + 5 =$ **e** $3 \times (9 \div 3) =$ **f** $5 + (4 \times 2) =$

g $(5 + 3) \div 2 =$ **h** $(5 \div 1) \times 4 =$ **i** $(7 - 4) \times (1 + 4) =$

j $(7 + 5) \div (6 - 3) =$

3 Copy each of these and then put in brackets to make each sum true.

a $4 \times 5 - 1 = 16$ **b** $8 \div 2 + 4 = 8$ **c** $8 - 3 \times 4 = 20$ **d** $12 - 5 \times 2 = 2$

e $3 \times 3 + 2 = 15$ **f** $12 \div 2 + 1 = 4$ **g** $9 \times 6 \div 3 = 18$ **h** $20 - 8 + 5 = 7$

i $6 + 4 \div 2 = 5$ **j** $16 \div 4 \div 2 = 8$

4 Put any of +, –, ×, ÷ or ( ) in each sum to make it true.

a $2 \quad 5 \quad 10 = 0$ **b** $10 \quad 2 \quad 5 = 1$ **c** $10 \quad 5 \quad 2 = 3$ **d** $10 \quad 2 \quad 5 = 4$

e $10 \quad 5 \quad 2 = 7$ **f** $5 \quad 10 \quad 2 = 10$ **g** $10 \quad 5 \quad 2 = 13$ **h** $5 \quad 10 \quad 2 = 17$

i $10 \quad 2 \quad 5 = 20$ **j** $5 \quad 10 \quad 2 = 25$

5 Amanda worked out $3 + 4 \times 5$ and got the answer 35. Andrew worked out $3 + 4 \times 5$ and got the answer 23. Explain why they got different answers.

HOMEWORK 5D

1 Write the value of each underlined digit.

a $5\underline{7}6$ **b** $37\underline{4}$ **c** $\underline{6}89$ **d** $\underline{4}785$ **e** $300\underline{7}$

f $7\underline{6}08$ **g** $354\underline{2}$ **h** $1\underline{2}\,745$ **i** $\underline{8}7\,409$ **j** $\underline{7}\,777\,777$

2 Write each of the following using just words.

a 7245 **b** 9072 **c** 29 450 **d** 2 760 000

3 Write each of the following using digits only.

a Eight thousand and five hundred **b** Forty two thousand and forty two

c Six million **d** Five million and five.

4 Write these numbers in order, putting the **smallest** first.

a 31, 20, 14, 22, 8, 25, 30, 12

b 159, 155, 176, 167, 170, 168, 151, 172

c 2100, 2070, 2002, 1990, 2010, 1998, 2000, 2092

5 Write these numbers in order, putting the **largest** first.

a 49, 62, 75, 57, 50, 72
b 988, 1052, 999, 1010, 980, 1007
c 4567, 4765, 4675, 4576, 4657, 4756

6 Using each of the digits 7, 8 and 9 only once in each number:

a write as many three-digit numbers as you can.
b which of your numbers is the smallest?
c which of your numbers is the largest?

7 Write down in order of size, largest first, all the two-digit numbers that can be made using 2, 4 and 6. (Each digit can be repeated.)

8 Copy each of these sentences, writing the numbers in words.

a The diameter of the Earth at the equator is 12 756 kilometres.
b The Moon is approximately 238 000 miles from the Earth.
c The greatest distance of the Earth from the Sun is 94 600 000 miles.

HOMEWORK 5E

1 Round off each of these numbers to the nearest 10.

a 34 **b** 67 **c** 23 **d** 49 **e** 55
f 11 **g** 95 **h** 123 **i** 109 **j** 125

2 Round off each of these numbers to the nearest 100.

a 231 **b** 389 **c** 410 **d** 777 **e** 850
f 117 **g** 585 **h** 250 **i** 975 **j** 1245

3 Round off each of these numbers to the nearest 1000.

a 2176 **b** 3800 **c** 6760 **d** 4455 **e** 1204
f 6782 **g** 5500 **h** 8808 **i** 1500 **j** 9999

4 Give these bus journey times to the nearest 5 minutes.

a 16 minutes **b** 28 minutes **c** 34 minutes **d** 42 minutes
e $23\frac{1}{2}$ minutes **f** $17\frac{1}{2}$ minutes

5 The selling prices of five houses in a village is as follows:

FOR SALE	FOR SALE	FOR SALE	FOR SALE	FOR SALE
£8400	**£12 900**	**£45 300**	**£75 550**	**£99 500**

Give the prices to the nearest thousand pounds.

6 Mark knows that he has £240 in his savings account to the nearest ten pounds.

a What is the smallest amount that he could have?
b What is the greatest amount that he could have?

7 The size of a crowd at an open air pop festival was reported to be 8000 to the nearest thousand.

a What is the lowest number that the crowd could be?
b What is the largest number that the crowd could be?

HOMEWORK 5F

1 Copy and work out each of these additions.

a	**b**	**c**	**d**	**e**
75	245	307	4158	4289
+ 23	+ 156	+ 293	+ 3951	532
				+ 96

2 Complete each of these additions.

a 25 + 89 + 12 **b** 211 + 385 + 46 **c** 125 + 88 + 720
d 478 + 207 + 300 **e** 1275 + 3245 + 524

3 Copy and complete each of these subtractions.

a	**b**	**c**	**d**	**e**
354	651	785	450	5421
–120	–128	–207	–178	–2568

4 Complete each of these subtractions.

a 386 – 296 **b** 709 – 518 **c** 452 – 386
d 800 – 258 **e** 7208 – 1564

5 Copy each of these and fill in the missing digits.

a	**b**	**c**	**d**
4 5	□7	3□4	□□□
+3□	+4□	+2 8 6	+ 2 8 7
□7	9 2	□4□	5 5 5

6 Copy each of these and fill in the missing digits.

a	**b**	**c**	**d**
7 5	3 2□	5 8 3	□□□
– 1□	–1□ 4	–□□□	– 2 4 8
□3	1 8 2	1 3 5	3 7 4

HOMEWORK 5G

1 Copy and work out each of the following.

a	**b**	**c**	**d**	**e**
24	38	124	408	359
× 3	× 4	× 5	× 6	× 8

2 Calculate each of these multiplications.

a 21×5 **b** 37×7 **c** 203×9 **d** 4×876 **e** 6×3214

3 By doing a suitable multiplication, answer each of these questions.

a How many people could seven 55-seater coaches hold?
b Adam buys seven postcards at 23p each. How much does he spend in pounds?
c Nails are packed in boxes of 144. How many nails are there in five boxes?
d Eight people book a holiday, costing £284 each. What is the total cost?
e How many yards are there in six miles if there are 1760 yards in a mile?

4 Calculate each of these divisions.

a $684 \div 2$ **b** $525 \div 3$ **c** $804 \div 4$ **d** $7260 \div 5$ **e** $2560 \div 8$

5 By doing a suitable division, answer each of these questions.

a In a school there are 288 students in eight forms in Year 10. If there are the same number of students in each form, how many students are there in each one?
b Phil jogs seven miles every morning. How many days will it take him to cover a total distance of 441 miles?

c In a supermarket cans of cola are sold in packs of six. If there are 750 cans on the shelf, how many packs are there?

d Sandra's wages for a month were £2060. Assuming there are four weeks in a month, how much does she earn in a week?

e Tickets for a charity disco were sold at £5 each? How many people bought tickets if the total sales were £1710?

6 Fractions

HOMEWORK 6A

1 What fraction is shaded in each of these diagrams?

a 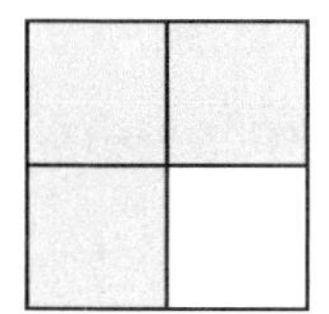

b 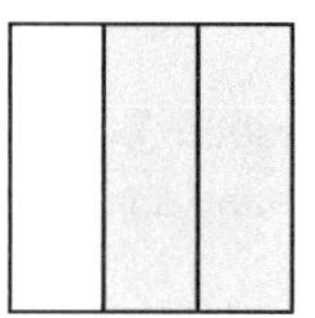

c 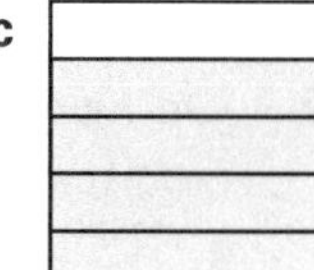

d 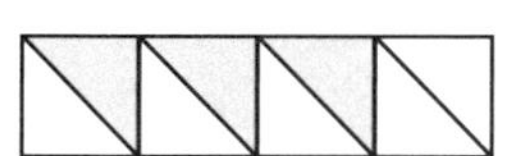

e 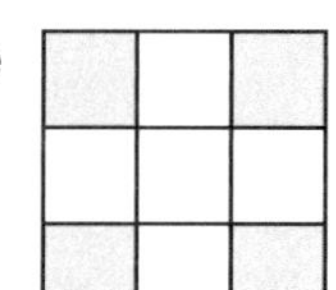

f

g 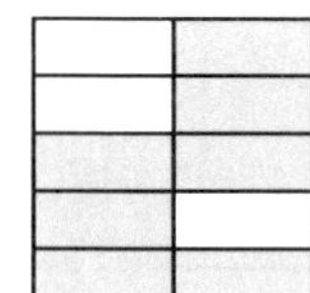

h 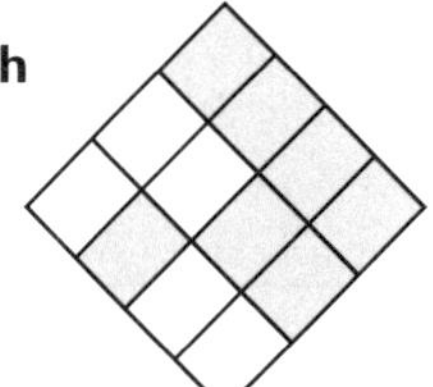

2 Draw diagrams as in Question **1** to show these fractions.

a $\frac{1}{3}$ **b** $\frac{3}{5}$ **c** $\frac{7}{10}$ **d** $\frac{5}{8}$ **e** $\frac{7}{9}$

f $\frac{3}{7}$ **g** $\frac{5}{12}$ **h** $\frac{7}{15}$

HOMEWORK 6B

Example **a** $\frac{5}{12} + \frac{4}{12} = \frac{9}{12}$

b $\frac{7}{10} - \frac{3}{10} = \frac{4}{10}$

1 Calculate each of the following.

a $\frac{1}{4} + \frac{1}{4}$ **b** $\frac{2}{5} + \frac{1}{5}$ **c** $\frac{3}{7} + \frac{2}{7}$ **d** $\frac{5}{8} + \frac{1}{8}$ **e** $\frac{3}{6} + \frac{2}{6}$

f $\frac{4}{9} + \frac{4}{9}$ **g** $\frac{3}{10} + \frac{4}{10}$ **h** $\frac{2}{5} + \frac{2}{5}$ **i** $\frac{4}{12} + \frac{1}{12}$ **j** $\frac{5}{20} + \frac{7}{20}$

2 Calculate each of the following.

a $\frac{4}{5} - \frac{2}{5}$ **b** $\frac{5}{8} - \frac{1}{8}$ **c** $\frac{6}{7} - \frac{2}{7}$ **d** $\frac{8}{10} - \frac{3}{10}$ **e** $\frac{5}{6} - \frac{3}{6}$

f $\frac{7}{9} - \frac{3}{9}$ **g** $\frac{7}{8} - \frac{1}{8}$ **h** $\frac{4}{9} - \frac{2}{9}$ **i** $\frac{7}{12} - \frac{5}{12}$ **j** $\frac{11}{20} - \frac{3}{20}$

3 **a** Draw two diagrams to show $\frac{4}{8}$ and $\frac{2}{8}$.

b Show on your diagrams that $\frac{4}{8} = \frac{1}{2}$ and $\frac{2}{8} = \frac{1}{4}$.

c Use the above information to write down the answers to each of the following.

i $\frac{1}{2} + \frac{1}{8}$ **ii** $\frac{1}{2} + \frac{3}{8}$ **iii** $\frac{1}{4} + \frac{1}{8}$ **iv** $\frac{3}{4} + \frac{1}{8}$ **v** $\frac{1}{2} - \frac{1}{8}$ **vi** $\frac{1}{2} - \frac{3}{8}$ **vii** $\frac{1}{4} - \frac{1}{8}$ **viii** $\frac{3}{4} - \frac{1}{8}$

HOMEWORK 6C

1 Copy the diagram and use it to write down each of these fractions as tenths.

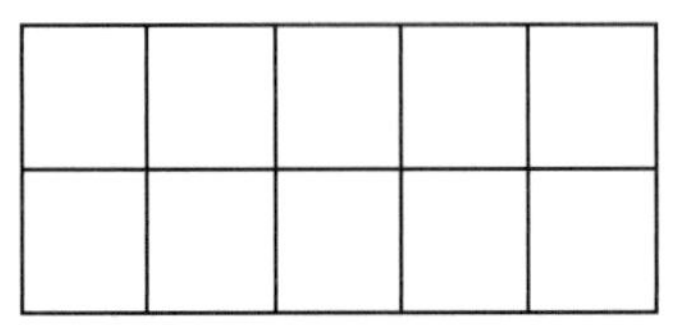

a $\frac{1}{2}$ **b** $\frac{1}{5}$ **c** $\frac{2}{5}$ **d** $\frac{3}{5}$ **e** $\frac{4}{5}$

2 Use your answers to Question **1** to write down the answer to each of the following. Each answer will be so many tenths.

a $\frac{1}{2} + \frac{1}{5}$ **b** $\frac{1}{2} + \frac{3}{10}$ **c** $\frac{2}{5} + \frac{1}{10}$ **d** $\frac{1}{5} + \frac{7}{10}$ **e** $\frac{1}{2} - \frac{2}{5}$ **f** $\frac{9}{10} - \frac{3}{5}$

3 Copy the diagram and use it to write down each of these fractions as twelfths.

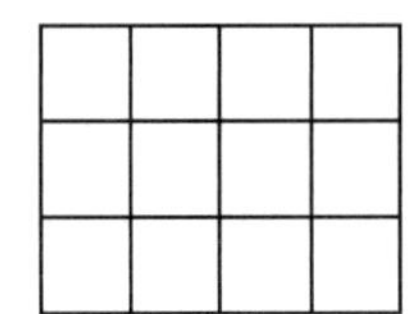

a $\frac{1}{2}$ **b** $\frac{1}{4}$ **c** $\frac{1}{3}$ **d** $\frac{3}{4}$ **e** $\frac{2}{3}$

4 Use your answers to Question **3** to write down the answer to each of the following. Each answer will be so many twelfths.

a $\frac{1}{2} + \frac{1}{3}$ **b** $\frac{1}{4} + \frac{1}{3}$ **c** $\frac{2}{3} + \frac{1}{4}$ **d** $\frac{1}{4} + \frac{7}{12}$ **e** $\frac{5}{12} + \frac{1}{2}$

f $\frac{2}{3} - \frac{1}{2}$ **g** $\frac{3}{4} - \frac{1}{3}$ **h** $\frac{1}{2} - \frac{1}{12}$ **i** $\frac{3}{4} - \frac{5}{12}$ **j** $\frac{2}{3} - \frac{7}{12}$

HOMEWORK 6D

Example 1 Show $\frac{2}{3}$ is equivalent to $\frac{8}{12}$.

$$\frac{2}{3} \Rightarrow \frac{\times 4}{\times 4} = \frac{8}{12}$$

Example 2 Cancel down $\frac{15}{20}$ to its simplest form.

$$\frac{15}{20} \Rightarrow \frac{\div 5}{\div 5} = \frac{3}{4}$$

Example 3 Put these fractions in order with the smallest first: $\frac{5}{6}$ $\frac{2}{3}$ $\frac{3}{4}$.

By equivalent fractions $\frac{5}{6} = \frac{10}{12}$ $\frac{2}{3} = \frac{8}{12}$ $\frac{3}{4} = \frac{9}{12}$

Putting in order: $\frac{2}{3}$ $\frac{3}{4}$ $\frac{5}{6}$

1 Copy and complete each of these statements.

a $\frac{1}{4} \Rightarrow \frac{\times 3}{\times 3} = \frac{}{12}$ **b** $\frac{3}{5} \Rightarrow \frac{\times 4}{\times 4} = \frac{}{20}$ **c** $\frac{5}{8} \Rightarrow \frac{\times 2}{\times 2} = \frac{}{16}$ **d** $\frac{4}{7} \Rightarrow \frac{\times 3}{\times 3} = \frac{12}{}$

e $\frac{2}{3} \Rightarrow \frac{\times}{\times 5} = \frac{}{15}$ **f** $\frac{5}{9} \Rightarrow \frac{\times}{\times 2} = \frac{}{18}$ **g** $\frac{6}{7} \Rightarrow \frac{\times}{\times} = \frac{}{35}$ **h** $\frac{1}{10} \Rightarrow \frac{\times}{\times} = \frac{}{40}$

2 Copy and complete each of these statements.

a $\frac{1}{4} = \frac{2}{8} = \frac{}{12} = \frac{4}{} = \frac{}{20} = \frac{6}{24}$ **b** $\frac{2}{3} = \frac{4}{6} = \frac{}{} = \frac{}{12} = \frac{10}{} = \frac{12}{18}$

c $\frac{4}{5} = \frac{}{10} = \frac{12}{} = \frac{}{20} = \frac{}{} = \frac{}{30}$ **d** $\frac{3}{10} = \frac{}{} = \frac{}{30} = \frac{}{} = \frac{}{50} = \frac{18}{}$

3 Copy and complete each of these statements.

a $\frac{6}{8} = \frac{6 \div 2}{8 \div 2} = \frac{}{}$ **b** $\frac{9}{12} = \frac{9 \div 3}{12 \div 3} = \frac{}{}$

c $\frac{15}{25} = \frac{15 \div 5}{25 \div} = \frac{}{}$ **d** $\frac{20}{70} = \frac{20 \div 10}{70 \div} = \frac{}{}$

4 Cancel down each of these fractions into their simplest form.

a $\frac{4}{10}$ **b** $\frac{3}{12}$ **c** $\frac{5}{25}$ **d** $\frac{6}{15}$ **e** $\frac{8}{12}$
f $\frac{10}{30}$ **g** $\frac{12}{20}$ **h** $\frac{16}{24}$ **i** $\frac{30}{50}$ **j** $\frac{42}{49}$

5 Put the following fractions in order with the **smallest** first.

a $\frac{1}{3}, \frac{1}{2}, \frac{1}{4}$ **b** $\frac{3}{4}, \frac{3}{8}, \frac{1}{2}$ **c** $\frac{5}{6}, \frac{2}{3}, \frac{7}{12}$ **d** $\frac{2}{5}, \frac{3}{10}, \frac{1}{4}$

HOMEWORK 6E

1 Change each of these top-heavy fractions into a mixed number.

a $\frac{5}{2}$ **b** $\frac{5}{3}$ **c** $\frac{7}{4}$ **d** $\frac{11}{3}$ **e** $\frac{9}{2}$ **f** $\frac{13}{4}$
g $\frac{11}{5}$ **h** $\frac{10}{4}$ **i** $\frac{14}{6}$ **j** $\frac{17}{8}$ **k** $\frac{17}{10}$ **l** $\frac{26}{8}$
m $\frac{12}{4}$ **n** $\frac{20}{5}$ **o** $\frac{60}{10}$

2 Change each of these mixed numbers into a top-heavy fraction.

a $1\frac{1}{2}$ **b** $2\frac{1}{4}$ **c** $2\frac{1}{3}$ **d** $4\frac{1}{2}$ **e** $3\frac{2}{3}$ **f** $1\frac{3}{4}$
g $2\frac{1}{5}$ **h** $2\frac{3}{8}$ **i** $3\frac{2}{5}$ **j** $4\frac{3}{5}$ **k** $5\frac{3}{8}$ **l** $4\frac{3}{7}$
m $5\frac{4}{9}$ **n** $4\frac{5}{12}$ **o** $7\frac{7}{10}$

HOMEWORK 6F

Example 1 $\frac{5}{9} + \frac{7}{9} = \frac{12}{9} = \frac{4}{3} = 1\frac{1}{3}$ (Cancel down and change to a mixed number.)

Example 2 $\frac{2}{3} + \frac{1}{5} = \frac{10}{15} + \frac{3}{15} = \frac{13}{15}$ (Use equivalent fractions to make the denominators the same.)

1 Calculate each of these additions. Remember to cancel down.

a $\frac{3}{8} + \frac{1}{8}$ **b** $\frac{3}{10} + \frac{5}{10}$ **c** $\frac{5}{12} + \frac{1}{12}$ **d** $\frac{1}{9} + \frac{5}{9}$ **e** $\frac{2}{15} + \frac{7}{15}$

2 Calculate each of these additions. Remember to change into mixed numbers.

a $\frac{5}{8} + \frac{7}{8}$ **b** $\frac{3}{4} + \frac{3}{4}$ **c** $\frac{7}{10} + \frac{3}{10}$ **d** $\frac{5}{12} + \frac{11}{12}$ **e** $\frac{13}{20} + \frac{11}{20}$

3 Calculate each of these additions. Remember to use equivalent fractions.

a $\frac{1}{3} + \frac{1}{2}$ **b** $\frac{2}{5} + \frac{3}{10}$ **c** $\frac{1}{4} + \frac{5}{12}$ **d** $\frac{3}{5} + \frac{1}{4}$
e $\frac{3}{4} + \frac{2}{3}$ **f** $\frac{5}{6} + \frac{1}{2}$ **g** $1\frac{1}{2} + 2\frac{1}{4}$ **h** $1\frac{1}{3} + 2\frac{3}{4}$

4 Calculate each of these subtractions.

a $\frac{5}{8} - \frac{1}{8}$	**b** $\frac{7}{10} - \frac{3}{10}$	**c** $\frac{11}{12} - \frac{3}{4}$	**d** $\frac{2}{3} - \frac{1}{2}$
e $\frac{9}{10} - \frac{1}{5}$	**f** $1 - \frac{3}{5}$	**g** $3 - 1\frac{1}{4}$	**h** $4\frac{3}{4} - 1\frac{1}{3}$

HOMEWORK 6G

1 At a cricket match, $\frac{9}{10}$ of the crowd were men. What fraction of the crowd were women?

2 An iceberg shows $\frac{1}{9}$ of its mass above sea level. What fraction of it is below sea level?

3 A petrol gauge shows that a tank is $\frac{7}{12}$ full. What fraction of the tank is empty?

4 David spends $\frac{1}{4}$ of his pocket money on bus fares, $\frac{1}{3}$ on magazines and saves the rest. What fraction of his money does he save?

★**5** In a local election Mr Weeks received $\frac{2}{5}$ of the total votes, Ms Meenan received $\frac{1}{4}$ and Mr White received the remainder. What fraction of the total votes did Mr White receive?

★**6** On a certain day at a busy railway station, $\frac{7}{10}$ of the trains arriving were on time, $\frac{1}{6}$ were late by 10 minutes or less and the rest were late by more than 10 minutes. What fraction of the trains arrived late by more than 10 minutes?

HOMEWORK 6H

1 Calculate each of these.

a $\frac{1}{2} \times 20$	**b** $\frac{1}{3} \times 36$	**c** $\frac{1}{4} \times 24$	**d** $\frac{3}{4} \times 40$
e $\frac{2}{3} \times 15$	**f** $\frac{1}{5} \times 30$	**g** $\frac{3}{8} \times 16$	**h** $\frac{7}{10} \times 50$

2 Calculate each of these quantities.

a $\frac{1}{4}$ of £800	**b** $\frac{2}{3}$ of 60 kilograms	**c** $\frac{3}{4}$ of 200 metres
d $\frac{3}{8}$ of 48 gallons	**e** $\frac{4}{5}$ of 30 minutes	**f** $\frac{7}{10}$ of 120 miles

3 In each case, find out which is the smaller number.

a $\frac{1}{4}$ of 60 or $\frac{1}{2}$ of 40	**b** $\frac{1}{3}$ of 36 or $\frac{1}{5}$ of 50
c $\frac{2}{3}$ of 15 or $\frac{3}{4}$ of 12	**d** $\frac{5}{8}$ of 72 or $\frac{5}{6}$ of 60

4 $\frac{5}{9}$ of a class of 36 students are girls. How many boys are there in the class?

5 Mrs Wilson puts $\frac{3}{20}$ of her weekly wage into a pension scheme. How much does she put into the scheme if her wage one week is £320?

6 Mitchell spent one third of a day sleeping and one quarter at school. How many hours are left for doing other things?

★**7** A bush is 40 cm tall when planted in spring. Its height increases by $\frac{3}{10}$ during the summer.

a Find $\frac{3}{10}$ of 40 cm.

b Find the height of the bush at the end of the summer.

★**8** A travel agent has this sign in their window. Marion books a holiday for her family which would normally cost £800.

a How much does she save?

b How much does she pay for the holiday after the reduction?

⅕ OFF

all holiday prices for next year if booked before December

HOMEWORK 6I

Example $\frac{2}{3} \times \frac{1}{4} = \frac{2}{12} = \frac{1}{6}$ Multiply numerators and denominators and cancel if possible.

Work out each of these multiplications.

1 $\frac{1}{2} \times \frac{1}{2}$ **2** $\frac{1}{3} \times \frac{1}{5}$ **3** $\frac{1}{4} \times \frac{1}{3}$ **4** $\frac{3}{4} \times \frac{1}{2}$ **5** $\frac{1}{3} \times \frac{3}{5}$

6 $\frac{2}{3} \times \frac{1}{2}$ **7** $\frac{4}{5} \times \frac{1}{2}$ **8** $\frac{5}{6} \times \frac{1}{5}$ **9** $\frac{3}{8} \times \frac{2}{3}$ **10** $\frac{3}{10} \times \frac{5}{6}$

HOMEWORK 6J

Example Write £5 as a fraction of £20.

$\frac{5}{20} = \frac{1}{4}$ (Cancel down)

1 Write the first quantity as a fraction of the second.

a £2, £8 **b** 9 cm, 12 cm **c** 18 miles, 30 miles

d 200 g, 350 g **e** 20 seconds, 1 minute **f** 25p, £2

2 During a one hour TV programme, 10 minutes were devoted to adverts. What fraction of the time was given to adverts?

3 On an 80 mile car journey, 50 miles were driven on a motorway. What fraction of the journey was not driven on a motorway?

4 In a class, 24 students were right-handed and 6 students were left-handed. What fraction of the class were:

a right-handed **b** left-handed?

HOMEWORK 6K

1 Work out each of these fractions as a decimal. Give them as terminating decimals or recurring decimals as appropriate.

a $\frac{3}{4}$ **b** $\frac{1}{15}$ **c** $\frac{1}{25}$ **d** $\frac{1}{11}$ **e** $\frac{1}{20}$

2 There are several patterns to be found in recurring decimals. For example,

$\frac{1}{13} = 0.076923076923076923076923\ldots$, $\frac{2}{13} = 0.153846153846153846153846\ldots$,

$\frac{3}{13} = 0.230769230769230769230769\ldots$ and so on.

a Write down the decimals for $\frac{4}{13}, \frac{5}{13}, \frac{6}{13}, \frac{7}{13}, \frac{8}{13}, \frac{9}{13}, \frac{10}{13}, \frac{11}{13}, \frac{12}{13}$ to 24 decimal places.

b What do you notice?

3 Write each of these fractions as a decimal. Use this to write the list in order of size, smallest first.

$\frac{2}{9}$ $\frac{1}{5}$ $\frac{23}{100}$ $\frac{2}{7}$ $\frac{3}{11}$

4 Convert each of these terminating decimals to a fraction in its simplest form.

a 0.57 **b** 0.275 **c** 0.85 **d** 0.06 **e** 3.65

5 Use a calculator to work out the reciprocal of each of the following.

a 4 **b** 8 **c** 32 **d** 40 **e** 100

6 Write down the reciprocal of each of the following fractions.

a $\frac{2}{3}$ **b** $\frac{5}{8}$ **c** $\frac{9}{10}$ **d** $\frac{7}{12}$ **e** $\frac{17}{20}$

7 Negative numbers

HOMEWORK 7A

Copy and complete each of the following.

1 If +£20 means a profit of twenty pounds, then .... means a loss of twenty pounds.

2 If –£10 means a loss of ten pounds, then +£10 means a .... of ten pounds.

3 If +500 m means 500 metres above sea level, then .... means 500 metres below sea level.

4 If –1000 m means one thousand metres below sea level, then +1000 m means one thousand metres .... sea level.

5 If +7 °C means seven degrees above freezing point, then .... means seven degrees below freezing point.

6 If +1 °C means 1 °C above freezing point, then .... means 1 °C below freezing point.

7 If –15 °C means fifteen degrees below freezing point, then +15 °C means fifteen degrees .... freezing point.

8 If –5000 m means five thousand miles south of the equator, then +5000 m means five thousand miles .... of the equator.

9 If a car moving forwards at 25 mph is represented by +25 mph, then a car moving backwards at 10 mph is represented by ....

10 In multi-storey car park, the sixth floor above ground level is represented by +6. So, the third floor below ground level is represented by ....

HOMEWORK 7B

Use the number line to answer Questions **1** and **2**.

–7 –6 –5 –4 –3 –2 –1 0 1 2 3 4 5 6 7

negative **positive**

1 Complete each of the following by putting a suitable number in the box.

a □ is smaller than 6 **b** □ is smaller than 2
c □ is smaller than –1 **d** □ is smaller than –6
e –4 is smaller than □ **f** –7 is smaller than □
g 5 is smaller than □ **h** 4 is smaller than □
i □ is smaller than 0 **j** –1 is smaller than □

2 Complete each of the following by putting a suitable number in the box.

a □ is bigger than –6 **b** □ is bigger than 4
c □ is bigger than –2 **d** □ is bigger than 0
e –3 is bigger than □ **f** –5 is bigger than □
g 1 is bigger than □ **h** –1 is bigger than □
i □ is bigger than –4 **j** 3 is bigger than □

3 In each case below, put the correct symbol, either < or >, in the box.

Reminder: The inequality signs: < means 'is less than' and > means 'is greater than'.

a	2 □ 6	**b**	–1 □ –7	**c**	–5 □ 1	**d**	5 □ 9
e	–8 □ 2	**f**	–14 □ –10	**g**	–11 □ 0	**h**	–9 □ –12
i	8 □ –3	**j**	0 □ –8				

4 Copy these number lines and fill in the missing numbers on each line.

a

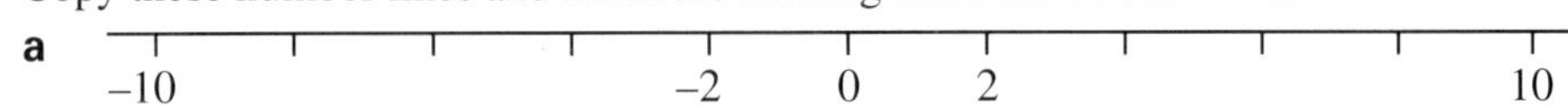

b

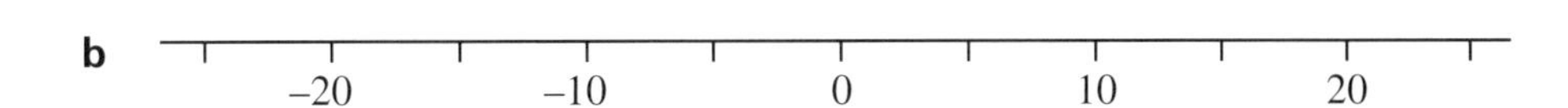

c

–20 0 20

d

–100 0 200

e

–100 0 100

HOMEWORK 7C

Example 1 –3 + 5 = 2

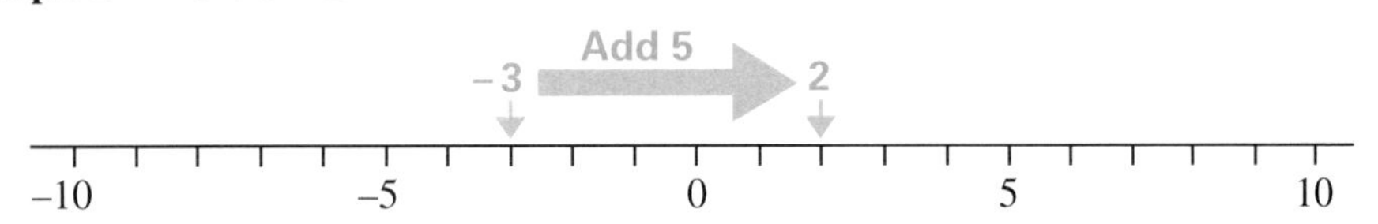

Example 2 –4 – 5 = –9

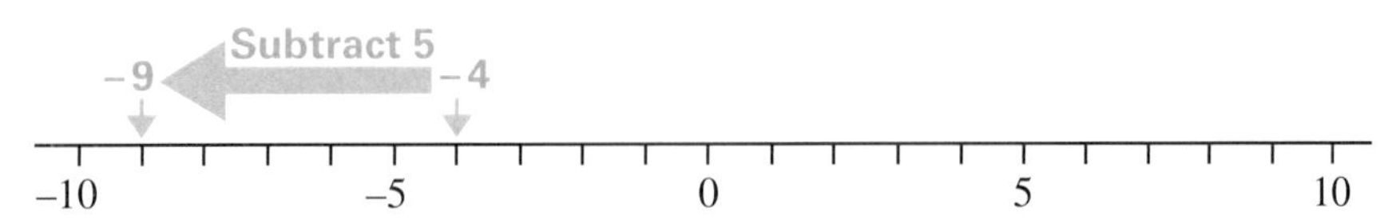

1 Use the number line to find the answer to each of the following.

a	–2 + 5 =	**b**	–4 + 6 =	**c**	–3 + 4 =	**d**	–1 + 5 =
e	–6 + 8 =	**f**	–5 + 10 =	**g**	–2 + 2 =	**h**	–4 + 4 =
i	4 – 5 =	**j**	6 – 8 =	**k**	3 – 7 =	**l**	5 – 9 =
m	–5 + 3 =	**n**	–2 + 1 =	**o**	–10 + 6 =	**p**	–8 + 6 =
q	–2 – 7 =	**r**	–1 – 5 =	**s**	–3 – 7 =	**t**	–5 – 5 =

2 Answer each of the following without the help of the number scale.

a	15 – 19 =	**b**	3 – 17 =	**c**	–2 – 10 =	**d**	–12 + 7 =
e	–15 + 9 =	**f**	10 – 20 =	**g**	–10 – 12 =	**h**	–15 – 20 =
i	23 – 30 =						

3 Work out each of the following.

a	1 + 3 – 5 =	**b**	–4 + 8 – 2 =	**c**	–6 + 3 – 4 =	**d**	–3 – 5 + 4 =
e	–1 – 1 + 5 =	**f**	–7 + 5 + 8 =	**g**	–3 – 4 + 7 =	**h**	1 – 3 – 6 =
i	–5 – 3 – 2 =						

HOMEWORK 7D

Example 1 $4 - (-2) = 4 + 2 = 6$

Example 2 $3 + (-5) = 3 - 5 = -2$

1 Answer each of the following. Check your answers on a calculator.

a $2 + (-5) =$ **b** $6 - (-3) =$ **c** $3 + (-5) =$ **d** $8 - (-2) =$
e $-6 + (-2) =$ **f** $-5 + (-2) =$ **g** $-2 - (-5) =$ **h** $-7 - (-1) =$
i $-2 - (-2) =$

2 Write down the answer to each of the following, then check your answers on a calculator.

a $-13 - 5 =$ **b** $-12 - 8 =$ **c** $-25 + 6 =$ **d** $6 - 14 =$
e $25 - -3 =$ **f** $13 - -8 =$ **g** $-4 + -15 =$ **h** $-13 + -7 =$
i $-12 + -9 =$ **j** $-16 + -12 =$

3 The temperature at midday was 5 °C. Find the temperature at midnight if it fell by:

a 1 °C **b** 5 °C **c** 6 °C **d** 8 °C **e** 12 °C.

4 What is the difference between the following temperatures?

a 4 °C and 6 °C **b** –2 °C and 4 °C **c** –3 °C and –6 °C

5 Rewrite the following lists, putting the numbers in order of size, smallest first.

a 2 –5 3 –6 –3 8 –1 1
b 4 –8 5 –10 –5 0 6 –12

★**6** You have the following cards.

5	3	2	–1	–3	–6

a Which other card should you choose to make the answer to the following sum as large as possible? What is the answer?

5 + ☐ = ☐

b Which other card should you choose to make the answer to part **a** as small as possible? What is the answer?

c Which other card should you choose to make the answer to the following sum as large as possible? What is the answer?

5 – ☐ = ☐

d Which other card should you choose to make the answer to part **c** as small as possible? What is the answer?

e Which two cards should you choose to make the answer to an addition sum zero?

HOMEWORK 7E

1 Find the next three numbers in each sequence.

a 6, 4, 2, 0, …, …, … **b** 8, 5, 2, –1, …, …, …
c –20, –15, –10, –5, …, …, … **d** 10, 9, 7, 4, …, …, …
e $-12, -10\frac{1}{2}, -9, -7\frac{1}{2}$, …, …, …

2 The deep freeze compartment in a refrigerator should be set at –14 °C, but in error is set to –6 °C. What is the difference between the two settings?

3 At 5am, the temperature on a thermometer outside Brian's house was –4 °C. By midday, the temperature had risen by 10 °C.

a What was the temperature at midday?

By midnight, the temperature had fallen to –9 °C.

b What was the fall in temperature from midday to midnight?

★**4** The table shows the recorded highest and lowest temperatures in five cities during one year.

	London	New York	Athens	Beijing	Nairobi
Highest temperature (°C)	30	28	36	31	29
Lowest temperature (°C)	–5	–8	5	–10	11

a Which city had the highest temperature?

b Which city had the largest difference in temperature and by how many degrees?

c Which city had the smallest difference in temperature and by how many degrees?

★**5** This is a two step function machine. Use the function machine to complete the table.

Number in ➡ **Add 2** ➡ **Subtract 5** ➡ **Number out**

Number in	Number out
10	
2	
–1	
	0
	–8

8 More about number

HOMEWORK 8A

1 Write out the first five multiples of:

a 4 **b** 6 **c** 8 **d** 12 **e** 15.

Remember: the first multiple is the number itself.

2 From the list of numbers below

28 19 36 43 64 53 77 66 56 60 15 29 61 45 51

write down those that are:

a multiples of 4 **b** multiples of 5 **c** multiples of 8 **d** multiples of 11.

3 Use your calculator to see which of the numbers below are:

a multiples of 7 **b** multiples of 9 **c** multiples of 12.

225 252 361 297 162 363 161 289 224 205 312 378 315 182 369

4 Find the biggest number smaller than 200 that is:

a a multiple of 2 **b** a multiple of 4 **c** a multiple of 5 **d** a multiple of 8

e a multiple of 9.

5 Find the smallest number that is a multiple of 3 and bigger than:

a 10 **b** 100 **c** 1000 **d** 10 000 **e** 1 000 000 000.

HOMEWORK 8B

Example Find the factors of 32.

Look for the pairs of numbers which make 32 when multiplied together. These are

$1 \times 32 = 32$, $2 \times 16 = 32$ and $4 \times 8 = 32$. So the factors of 32 are 1, 2, 4, 8, 16, 32.

1 What are the factors of each of these numbers?

a 12 **b** 13 **c** 15 **d** 20 **e** 22
f 36 **g** 42 **h** 48 **i** 49 **j** 50

2 Use your calculator to find the factors of each of these numbers.

a 100 **b** 111 **c** 125 **d** 132 **e** 140

3 All the numbers in **a** to **j** are divisible by 11. Use your calculator to divide each one by 11 and then write down the answer. What do you notice?

a 143 **b** 253 **c** 275 **d** 363 **e** 462
f 484 **g** 561 **h** 583 **i** 792 **j** 891

HOMEWORK 8C

1 Write down all the prime numbers less than 40.

2 Which of these numbers are prime?

43 47 49 51 54 57 59 61 65 67

3 This is a number pattern to generate odd numbers.

Line 1 $2 - 1 = 1$
Line 2 $2 \times 2 - 1 = 3$
Line 3 $2 \times 2 \times 2 - 1 = 7$

a Work out the next three lines of the pattern.
b Which lines have answers that are prime numbers?

4 Write down the first ten square numbers.

5 Here is another number pattern.

$2 \times 0 + 1 = 1$
$3 \times 1 + 1 = 4$
$4 \times 2 + 1 = 9$

a Write down the next three lines in the pattern.
b Describe what you notice about the answers to each line of the pattern.

6 Write down the answer to each of the following. You will need to use your calculator.

a 5^2 **b** 15^2 **c** 25^2 **d** 35^2 **e** 45^2
f 55^2 **g** 65^2 **h** 75^2 **i** 85^2 **j** 95^2

Describe any pattern you notice.

HOMEWORK 8D

1 Write down the first five multiples of:

a 5 **b** 7 **c** 16 **d** 25 **e** 30.

Remember: the first multiple is the number itself.

2 Write down the first three numbers that are multiples of both:

a 2 and 5 **b** 3 and 4 **c** 5 and 6 **d** 4 and 6 **e** 8 and 10.

3 Write down all the factors of each of these numbers.

a 18 **b** 25 **c** 28 **d** 35 **e** 40

4 From the list of numbers below

4 6 7 10 13 16 21 23 25 28 34 37 40 49 50

write down those that are:

a prime numbers **b** square numbers **c** triangle numbers.

5 In a prize draw, raffle tickets are numbered from 1 to 100.
A prize is given if a ticket drawn is a multiple of 10 or a multiple of 15.
Which ticket holders will receive two prizes?

6 Here is a number pattern using square numbers.

$1^2 - 0^2 = 1$
$2^2 - 1^2 = 3$
$3^2 - 2^2 = 5$
$4^2 - 3^2 = 7$

a Write down the next three lines in the pattern.

b What do you think is the answer to $21^2 - 20^2$?
Explain your answer.

HOMEWORK 8E

1 Write down the positive square root of each of these numbers.

a 64 **b** 25 **c** 49 **d** 81 **e** 16
f 36 **g** 100 **h** 121 **i** 144 **j** 400

2 Write down the answer to each of the following. You will need to use your calculator.

a $\sqrt{225}$ **b** $\sqrt{289}$ **c** $\sqrt{441}$ **d** $\sqrt{625}$ **e** $\sqrt{1089}$
f $\sqrt{1369}$ **g** $\sqrt{3136}$ **h** $\sqrt{6084}$ **i** $\sqrt{40\,804}$ **j** $\sqrt{110\,889}$

3 Here is a number pattern using square roots and square numbers.

$\sqrt{1} = 1$
$\sqrt{1} + \sqrt{4} = 3$
$\sqrt{1} + \sqrt{4} + \sqrt{9} = 6$

a Write down the next three lines in the pattern.

b Describe any pattern you notice in the answers.

HOMEWORK 8F

Example Work out 3^5.

$3^5 = 3 \times 3 \times 3 \times 3 \times 3 = 243$

1 Use your calculator to work out the value of each of the following.

a	2^3	**b**	4^3	**c**	7^3	**d**	10^3	**e**	12^3
f	3^4	**g**	10^4	**h**	2^5	**i**	10^6	**j**	2^8

2 Use your calculator to work out the answers to the following powers of 11.

a 11^2 **b** 11^3 **c** 11^4

Describe any patterns you notice in your answers.

Does your pattern work for other powers of 11? Give a reason for your answer.

★**3**

1	2	3	4	5	6	7	8	9
10	11	12	13	14	15	16	17	18
19	20	21	22	23	24	25	26	27
28	29	30	31	32	33	34	35	36

From the numbers above, write down:

a all the multiples of 7 **b** all the factors of 30

c all the prime numbers **d** the square of 6

e the square root of 25 **f** the cube of 3.

HOMEWORK 8G

1 Evaluate the following.

a	3.5×100	**b**	2.15×10	**c**	6.74×1000	**d**	4.63×10
e	30.145×10	**f**	78.56×1000	**g**	6.42×10^2	**h**	0.067×10
i	0.085×10^3	**j**	0.798×10^5	**k**	0.658×1000	**l**	215.3×10^2
m	0.889×10^6	**n**	352.147×10^2	**o**	37.2841×10^3	**p**	34.28×10^6

2 Evaluate the following.

a	$4538 \div 100$	**b**	$435 \div 10$	**c**	$76459 \div 1000$	**d**	$643.7 \div 10$
e	$4228.7 \div 100$	**f**	$278.4 \div 1000$	**g**	$246.5 \div 10^2$	**h**	$76.3 \div 10$
i	$76 \div 10^3$	**j**	$897 \div 10^5$	**k**	$86.5 \div 1000$	**l**	$1.5 \div 10^2$
m	$0.8799 \div 10^6$	**n**	$23.4 \div 10^2$	**o**	$7654 \div 10^3$	**p**	$73.2 \div 10^6$

3 Evaluate the following.

a	400×300	**b**	50×4000	**c**	70×200	**d**	30×700
e	$(30)^2$	**f**	$(50)^3$	**g**	$(200)^2$	**h**	40×150
i	70×200	**j**	60×5000	**k**	30×250	**l**	700×200

4 Evaluate the following.

a	$4000 \div 800$	**b**	$9000 \div 30$	**c**	$7000 \div 200$	**d**	$8000 \div 200$
e	$2100 \div 700$	**f**	$9000 \div 60$	**g**	$700 \div 50$	**h**	$3500 \div 70$
i	$3000 \div 500$	**j**	$30\,000 \div 2000$	**k**	$5600 \div 1400$	**l**	$6000 \div 30$

5 Evaluate the following.

a	7.3×10^2	**b**	3.29×10^5	**c**	7.94×10^3	**d**	6.8×10^7
e	$3.46 \div 10^2$	**f**	$5.07 \div 10^4$	**g**	$2.3 \div 10^4$	**h**	$0.89 \div 10^3$

HOMEWORK 8H

Example $2^2 \times 3 \times 5 = 4 \times 3 \times 5 = 60$

1 Copy and complete the following prime factor trees.

a

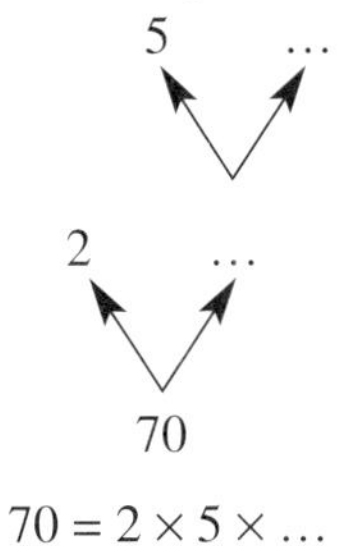

$70 = 2 \times 5 \times \ldots$

b

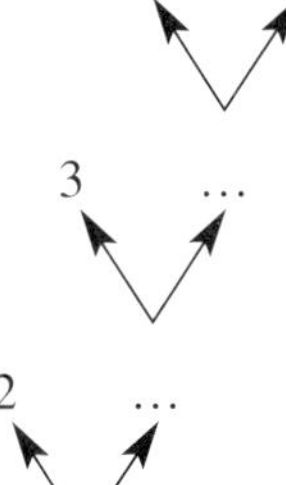

$90 = 2 \times 3 \times 3 \times \ldots$

c

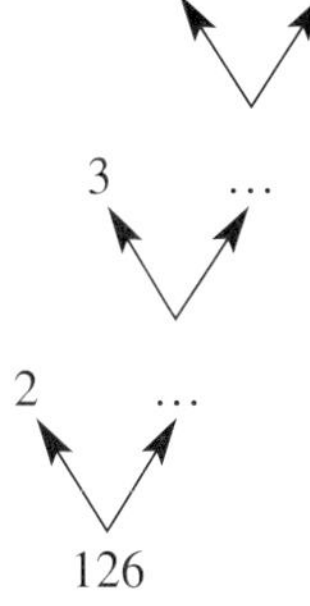

$126 = 2 \times 3 \times \ldots \times \ldots$

2 Write the following as numbers.

a $2^2 \times 3^2$ **b** $2 \times 3 \times 5^2$ **c** $3^2 \times 7$ **d** $2^3 \times 3 \times 5^2$ **e** $3^3 \times 5^2$

3 Write the following numbers as products of their prime factors.

a 24 **b** 36 **c** 75 **d** 84 **e** 99

HOMEWORK 8I

1 Find the lowest common multiple of these pairs of numbers.

a 3 and 4 **b** 6 and 8 **c** 9 and 12
d 10 and 12 **e** 14 and 21 **f** 20 and 24

2 Find the highest common factor of these pairs of numbers.

a 16 and 24 **b** 28 and 35 **c** 24 and 30
d 48 and 60 **e** 28 and 70 **f** 75 and 125

3 For each set of numbers, find **i** the lowest common multiple and **ii** the highest common factor.

a 2, 4 and 6 **b** 4, 6 and 8 **c** 8, 12 and 16 **d** 6, 12 and 15 **e** 20, 25 and 30

HOMEWORK 4J

1 Write each of the following as a single power of 7.

a $7^2 \times 7^3$ **b** $7^4 \times 7^5$ **c** 7×7^3 **d** $7^8 \times 7^2$ **e** $7^3 \times 7^4 \times 7^5$

2 Write each of the following as a single power of 4.

a $4^8 \div 4^3$ **b** $4^5 \div 4^2$ **c** $4^7 \div 4^5$ **d** $4^6 \div 4^5$ **e** $4^8 \times 4^4 \div 4^3$

3 Write each of the following as a single power of x.

a $x^2 \times x^3$ **b** $x^4 \times x^5$ **c** $x^6 \times x$ **d** $x^5 \times x^5$ **e** $x^3 \times x^2 \times x^4$

4 Write each of the following as a single power of y.

a $y^5 \div y^2$ **b** $y^8 \div y^3$ **c** $y^{10} \div y$ **d** $y^{12} \div y^4$ **e** $\frac{y^8 \times y^2}{y^3}$

9 Further number skills

HOMEWORK 9A

1 24×13 **2** 33×17 **3** 54×42 **4** 89×23 **5** 58×53

6 176×14 **7** 235×16 **8** 439×21 **9** 572×35 **10** 678×57

HOMEWORK 9B

1 $312 \div 13$ **2** $480 \div 15$ **3** $697 \div 17$ **4** $792 \div 22$ **5** $806 \div 26$

6 $532 \div 28$ **7** $736 \div 32$ **8** $595 \div 35$ **9** $948 \div 41$ **10** $950 \div 53$

HOMEWORK 9C

1 Wall tiles are packed in boxes of 16. Andy buys 24 packs to tile his bathroom. How many tiles does he buy altogether?

2 The organiser of a church fete requires 1000 coloured balloons. How many packets does she need to buy if there are 25 balloons in a packet?

3 A TV rental shop purchases 32 televisions at £112 each.

a Find the total cost of the televisions.

b Show how you could check your answer by estimation.

4 The annual subscription fee to join a Fishing Club is £42. The treasurer of the club has collected £1134 in fees. How many people have paid their subscription fee?

5 Mrs Woodhead saves £14 per week towards her bills. How much does she save in a year?

6 Sylvia has a part time job and is paid £18 for every day she works. Last year she worked for 148 days. How much was she paid for the year?

7 A coach firm charges £504 for 36 people to go Christmas shopping on a day trip to Calais. How much does each person pay if they share the cost equally between them?

8 A concert hall has 48 rows of seats with 32 seats in a row. What is the maximum capacity of the hall?

★**9** Allan is a market gardener and has 420 bulbs to plant. He plants them out in rows with 18 bulbs to a row. How many complete rows will there be?

★**10** A room measuring 6 metres by 8 metres is to be carpeted. The carpet costs £19 per square metre.

a Estimate the cost of the carpet.

b Calculate the exact cost of the carpet.

HOMEWORK 9D

Example 1 5.852 will round off to 5.85 to two decimal places

Example 2 7.156 will round off to 7.16 to two decimal places

Example 3 0.284 will round off to 0.3 to one decimal place

Example 4 15.3518 will round off to 15.4 to one decimal place

1 Round each of the following numbers to one decimal place.

a	3.73	**b**	8.69	**c**	5.34	**d**	18.75	**e**	0.423
f	26.288	**g**	3.755	**h**	10.056	**i**	11.08	**j**	12.041

2 Round each of the following numbers to two decimal places.

a	6.721	**b**	4.457	**c**	1.972	**d**	3.485	**e**	5.807
f	2.564	**g**	21.799	**h**	12.985	**i**	2.302	**j**	5.555

3 Round each of the following to the number of decimal places indicated.

a	4.572 (1 dp)	**b**	0.085 (2 dp)	**c**	5.7159 (3 dp)	**d**	4.558 (2 dp)
e	2.099 (2 dp)	**f**	0.7629 (3 dp)	**g**	7.124 (1 dp)	**h**	8.903 (2 dp)
i	23.7809 (3 dp)	**j**	0.99 (1 dp)				

4 Round each of the following to the nearest whole number.

a	6.7	**b**	9.3	**c**	2.8	**d**	7.5	**e**	8.38
f	2.82	**g**	2.18	**h**	1.55	**i**	5.252	**j**	3.999

HOMEWORK 9E

Example Work out 4.2 + 8 + 12.93. Set out the sum as follows:

Remember to keep the points in the same column.

$$\begin{array}{r} 4.20 \\ 8.00 \\ +\ 12.93 \\ \hline 25.13 \\ \hline {\scriptstyle 1\,1} \end{array}$$

1 Work out each of these.

a	7.3 + 2.6	**b**	15.7 + 5.6	**c**	33.5 + 6.8	**d**	8.5 + 4.82
e	3.26 + 4.5	**f**	2.75 + 9.84	**g**	24.5 + 6.3	**h**	8.4 + 12.8
i	13.75 + 8.5	**j**	7.08 + 0.7	**k**	7 + 2.96 + 3.1	**l**	8.5 + 7.36 + 12.1

2 Work out each of these.

a	5.8 – 3.4	**b**	7.3 – 2.8	**c**	4.6 – 2.7	**d**	9.7 – 4.7
e	8.35 – 4.24	**f**	9.74 – 3.81	**g**	9.04 – 5.72	**h**	3.62 – 1.85
i	6 – 3.3	**j**	8 – 7.4	**k**	12 – 3.2	**l**	7.2 – 4.72

HOMEWORK 9F

Example 1 $4.5 \times 3 =$

$$\begin{array}{r} 4.5 \\ \times\ \ 3 \\ \hline 13.5 \\ \hline {\scriptstyle 1} \end{array}$$

Example 2 $8.25 \div 5 =$

$$\begin{array}{r} 1.6\,5 \\ 5\,\overline{)\,8.{}^{3}2{}^{2}5} \end{array}$$

Example 3 $5.7 \div 2 =$

$$2 \overline{)5.{}^{1}7{}^{1}0} = 2.85$$

1 Evaluate each of these.

a 2.3×3 **b** 4.8×2 **c** 4.6×4 **d** 15.3×5 **e** 26.4×8

2 Evaluate each of these.

a 2.14×2 **b** 3.45×3 **c** 5.47×6 **d** 4.44×8 **e** 0.25×9

3 Evaluate each of these.

a $4.8 \div 2$ **b** $7.6 \div 4$ **c** $7.2 \div 3$ **d** $7.35 \div 5$ **e** $0.78 \div 6$

4 Evaluate each of these.

a $4.5 \div 2$ **b** $7.2 \div 5$ **c** $3.4 \div 4$ **d** $13.1 \div 5$ **e** $6.3 \div 8$

★**5** Crisps are sold in packs of six for £1.32 or packs of eight for £1.92. Which are better value?

★**6** Steve took his wife and three children on a day trip by train to London. The tickets were £26.60 for each adult and £12.85 for each child. How much did the tickets cost Steve altogether?

HOMEWORK 9G

Example Evaluate $4.27 \times 34 =$

```
   4.27
 ×   34
  17.08  (multiply by 4)
 128.10  (multiply by 3 and keep points in same column)
 145.18
```

1 Evaluate each of these.

a 3.12×14 **b** 5.24×15 **c** 1.36×22 **d** 7.53×25 **e** 27.1×32

2 Find the total cost of each of the following purchases.

a Twenty-four litres of petrol at £0.92 per litre

b Eighteen pints of milk at £0.32 per pint.

c Fourteen magazines at £2.25 per copy.

3 The table shows the exchange rate for various currencies

Currency	Exchange rate
Euro (€)	£1 = €1.42
American dollar ($)	£1 = $1.72
Swiss franc (F)	£1 = 2.24F

a Douglas changes £25 into euros. How many euros does he get?

b Martin changes £32 into dollars. How many dollars does he get?

c Pauline changes £45 into francs. How many francs does she get?

HOMEWORK 9H

To multiply one decimal number by another decimal number:

- First, do the whole calculation as if the decimal points were not there.
- Then, count the total number of decimal places in the two decimal numbers. This gives the number of decimal places in the answer.

Example Evaluate 3.42×0.2

Ignoring the decimal points gives the following calculation: $342 \times 2 = 684$

Now, 3.42 has 2 decimal places and 0.2 has 1 decimal place. So, the total number of decimal places in the answer is 3, which gives $3.42 \times 0.2 = 0.684$

1 Evaluate each of these.

a 2.3×0.2 **b** 5.2×0.3 **c** 4.6×0.4 **d** 0.2×0.3 **e** 0.4×0.7
f 0.5×0.5 **g** 12.6×0.6 **h** 7.2×0.7 **i** 1.4×1.2 **j** 2.6×1.5

2 For each of the following:

i estimate the answer by first rounding off each number to the nearest whole number.
ii calculate the exact answer, and then, by doing a subtraction, calculate how much out your answer to part **i** is.

a 3.7×2.4 **b** 4.8×3.1 **c** 5.1×4.2 **d** 6.5×2.5

HOMEWORK 9I

Example 1 Express 0.32 as a fraction.

$0.32 = \frac{32}{100}$. This cancels down to $\frac{8}{25}$.

Example 2 Express $\frac{3}{8}$ as a decimal. $\frac{3}{8} = 3 \div 8 = 0.375$.

$$8 \overline{)3.{}^{3}0{}^{6}0{}^{4}0} = 0.375$$

Notice how the extra zeros have been added.

1 Change each of these decimals to fractions, cancelling down where possible.

a 0.3 **b** 0.8 **c** 0.9 **d** 0.07 **e** 0.08
f 0.15 **g** 0.75 **h** 0.48 **i** 0. 32 **j** 0.27

2 Change each of these fractions to decimals.

a $\frac{1}{4}$ **b** $\frac{2}{5}$ **c** $\frac{7}{10}$ **d** $\frac{9}{20}$ **e** $\frac{7}{8}$

3 Put each of the following sets of numbers in order with the smallest first. It is easier to change the fractions into decimals first.

a $0.3, 0.2, \frac{2}{5}$ **b** $\frac{7}{10}, 0.8, 0.6$ **c** $0.4, \frac{1}{4}, 0.2$
d $\frac{3}{10}, 0.32, 0.29$ **e** $0.81, \frac{4}{5}, 0.78$

HOMEWORK 9J

1 Evaluate the following.

a $\frac{1}{2}+\frac{1}{5}$ **b** $\frac{1}{2}+\frac{1}{3}$ **c** $\frac{1}{3}+\frac{1}{10}$ **d** $\frac{3}{8}+\frac{1}{3}$
e $\frac{3}{4}+\frac{1}{5}$ **f** $\frac{1}{3}+\frac{2}{5}$ **g** $\frac{3}{5}+\frac{3}{8}$ **h** $\frac{1}{2}+\frac{2}{5}$

2 Evaluate the following.

a $\frac{1}{2}+\frac{1}{4}$ **b** $\frac{1}{3}+\frac{1}{6}$ **c** $\frac{3}{5}+\frac{1}{10}$ **d** $\frac{5}{8}+\frac{1}{4}$

3 Evaluate the following.

a $\frac{7}{8}-\frac{3}{4}$ **b** $\frac{4}{5}-\frac{1}{2}$ **c** $\frac{2}{3}-\frac{1}{5}$ **d** $\frac{3}{4}-\frac{2}{5}$

4 Evaluate the following.

a $\frac{5}{8}+\frac{3}{4}$ **b** $\frac{1}{2}+\frac{3}{5}$ **c** $\frac{5}{6}+\frac{1}{4}$ **d** $\frac{2}{3}+\frac{3}{4}$

5 Evaluate the following.

a $2\frac{1}{3}+1\frac{1}{4}$ **b** $3\frac{7}{10}+2\frac{3}{4}$ **c** $5\frac{3}{8}-2\frac{1}{3}$ **d** $4\frac{2}{5}-2\frac{5}{6}$

6 At a football club half of the players are English, a quarter are Scottish and one sixth are Italian. The rest are Irish. What fraction of players at the club are Irish?

7 On a firm's coach trip, half the people were employees, two fifths were partners of the employees. The rest were children. What fraction of the people were children?

8 Five eighths of the 35 000 crowd were male. How many females were in the crowd?

9 What is four fifths of sixty-five added to five sixths of fifty-four?

HOMEWORK 9K

1 Evaluate the following, leaving your answer in its simplest form.

a $\frac{1}{2}\times\frac{2}{3}$ **b** $\frac{3}{4}\times\frac{2}{5}$ **c** $\frac{3}{5}\times\frac{1}{2}$ **d** $\frac{3}{7}\times\frac{2}{3}$ **e** $\frac{2}{3}\times\frac{5}{6}$

f $\frac{1}{3}\times\frac{3}{5}$ **g** $\frac{2}{3}\times\frac{7}{10}$ **h** $\frac{3}{8}\times\frac{2}{5}$ **i** $\frac{4}{9}\times\frac{3}{8}$ **j** $\frac{4}{5}\times\frac{7}{16}$

2 Evaluate the following, leaving your answer as a mixed number where possible.

a $1\frac{1}{3}\times2\frac{1}{4}$ **b** $1\frac{3}{4}\times1\frac{1}{3}$ **c** $2\frac{1}{2}\times\frac{4}{5}$ **d** $1\frac{2}{3}\times1\frac{3}{10}$

e $3\frac{1}{4}\times1\frac{3}{5}$ **f** $2\frac{2}{3}\times1\frac{3}{4}$ **g** $3\frac{1}{2}\times1\frac{1}{6}$ **h** $7\frac{1}{2}\times1\frac{3}{5}$

3 Kris walked three quarters of the way along Carterknowle Road which is 3 km long. How far did Kris walk?

4 Jean ate one fifth of a cake, Les ate a half of what was left. Nick ate the rest. What fraction of the cake did Nick eat?

5 Billie made a cast that weighed five and three quarter kilograms. Four fifths of this weight is water. What is the weight of the water in Billie's cast?

6 Which is the smaller, $\frac{3}{4}$ of $5\frac{1}{3}$ or $\frac{2}{3}$ of $4\frac{2}{5}$?

7 I bought 24 bottles of lemonade, all containing $2\frac{3}{4}$ litres of lemonade. What is the total amount of lemonade I bought?

HOMEWORK 9L

1 Evaluate the following, leaving your answer as a mixed number where possible.

a $\frac{1}{5}\div\frac{1}{3}$ **b** $\frac{3}{5}\div\frac{3}{8}$ **c** $\frac{4}{5}\div\frac{2}{3}$ **d** $\frac{4}{7}\div\frac{8}{9}$

e $4\div1\frac{1}{2}$ **f** $5\div3\frac{2}{3}$ **g** $8\div1\frac{3}{4}$ **h** $6\div1\frac{1}{4}$

i $5\frac{1}{2}\div1\frac{1}{3}$ **j** $7\frac{1}{2}\div2\frac{2}{3}$ **k** $1\frac{1}{2}\div1\frac{1}{5}$ **l** $3\frac{1}{5}\div3\frac{3}{4}$

2 A pet shop has thirty-six kilograms of hamster food. Tom, who owns the shop, wants to pack this into bags, each containing three quarters of a kilogram. How many bags can he make in this way?

3 Bob is putting a fence down the side of his garden, it is to be 20 metres long. The fence comes in sections; each one is one and one third of a metre long. How many sections will Bob need to put the fence all the way down the one side of his garden?

4 An African Bullfrog can jump a distance of $1\frac{1}{4}$ metres in one hop. How many hops would it take an African Bullfrog to hop a distance of 100 metres?

5 Evaluate the following, leaving your answer as a mixed number wherever possible.

a $\frac{4}{5}\times\frac{1}{2}\times\frac{3}{8}$ **b** $\frac{3}{4}\times\frac{7}{10}\times\frac{5}{6}$ **c** $\frac{2}{3}\times\frac{5}{6}\times\frac{9}{10}$
d $1\frac{1}{4}\times\frac{2}{3}\div\frac{5}{6}$ **e** $\frac{5}{8}\times1\frac{1}{3}\div1\frac{1}{10}$ **f** $2\frac{1}{2}\times1\frac{1}{3}\div3\frac{1}{3}$

HOMEWORK 9M

1 Write down the answers to the following.

a	-2×4	**b**	-3×6	**c**	-5×7	**d**	-3×-4	**e**	-8×-2
f	$-14\div-2$	**g**	$-16\div-4$	**h**	$25\div-5$	**i**	$-16\div-8$	**j**	$-8\div-4$
k	3×-7	**l**	6×-3	**m**	7×-4	**n**	-3×-9	**o**	-7×-2
p	$28\div-4$	**q**	$12\div-3$	**r**	$-40\div8$	**s**	$-15\div-3$	**t**	$50\div-2$
u	-3×-8	**v**	$42\div-6$	**w**	7×-9	**x**	$-24\div-4$	**y**	-7×8

2 Write down the answers to the following.

a	$-2+4$	**b**	$-3+6$	**c**	$-5+7$	**d**	$-3+-4$	**e**	$-8+-2$
f	$-14--2$	**g**	$-16--4$	**h**	$25--5$	**i**	$-16--8$	**j**	$-8--4$
k	$3+-7$	**l**	$6+-3$	**m**	$7+-4$	**n**	$-3+-9$	**o**	$-7+-2$
p	$28--4$	**q**	$12--3$	**r**	$-40-8$	**s**	$-15--3$	**t**	$50--2$
u	$-3+-8$	**v**	$42--6$	**w**	$7+-9$	**x**	$-24--4$	**y**	$-7+8$

3 What number do you multiply –5 by to get the following?

a 25 **b** –30 **c** 50 **d** –100 **e** 75

HOMEWORK 9N

1 Round each of the following numbers to 1 significant figure.

a	46 313	**b**	57 123	**c**	30 569	**d**	94 558	**e**	85 299
f	54.26	**g**	85.18	**h**	27.09	**i**	96.432	**j**	167.77
k	0.5388	**l**	0.2823	**m**	0.005 84	**n**	0.047 85	**o**	0.000 876
p	9.9	**q**	89.5	**r**	90.78	**s**	199	**t**	999.99

2 What is the least and the greatest number of people that can be found in these towns?

Hellaby	population 900 (to 1 significant figure)
Hook	population 650 (to 2 significant figures)
Hundleton	population 1050 (to 3 significant figures)

3 Round each of the following numbers to 2 significant figures.

a	6725	**b**	35 724	**c**	68 522	**d**	41 689	**e**	27 308
f	6973	**g**	2174	**h**	958	**i**	439	**j**	327.6

4 Round each of the following to the number of significant figures (sf) indicated.

a	46 302 (1 sf)	**b**	6177 (2 sf)	**c**	89.67 (3 sf)	**d**	216.7 (2 sf)
e	7.78 (1 sf)	**f**	1.087 (2 sf)	**g**	729.9 (3 sf)	**h**	5821 (1 sf)
i	66.51 (2 sf)	**j**	5.986 (1 sf)	**k**	7.552 (1 sf)	**l**	9.7454 (3 sf)
m	25.76 (2 sf)	**n**	28.53 (1 sf)	**o**	869.89 (3 sf)	**p**	35.88 (1 sf)
q	0.084 71 (2 sf)	**r**	0.0099 (2 sf)	**s**	0.0809 (1 sf)	**t**	0.061 97 (3 sf)

HOMEWORK 9P

1 Find approximate answers to the following sums.

a	4324×6.71	**b**	6170×7.311	**c**	72.35×3.142
d	4709×3.81	**e**	$63.1 \times 4.18 \times 8.32$	**f**	$320 \times 6.95 \times 0.98$
g	$454 \div 89.3$	**h**	$26.8 \div 2.97$	**i**	$4964 \div 7.23$
j	$316 \div 3.87$	**k**	$2489 \div 48.58$	**l**	$63.94 \div 8.302$

2 Find the approximate monthly pay of the following people whose annual salary is

a Joy £47 200 **b** Amy £24 200 **c** Tom £19 135

3 Find the approximate annual pay of these brothers who earn:

a Trevor £570 a week **b** Brian £2728 a month

4 A litre of creosote will cover an area of about 6.8 m^2. Approximately how many litre cans will I need to buy to creosote a fence with a total surface area of 43 m^2?

★**5** A groundsman bought 350 kg of seed at a cost of £3.84 per kg. Find the approximate total cost of this seed.

6 A greengrocer sells a box of 250 apples for £47. Approximately how much did each apple sell for?

7 Keith runs about 15 km every day. Approximately how far does he run in:

a a week **b** a month **c** a year?

HOMEWORK 9Q

1 Round each of the following figures to a suitable degree of accuracy.

a Kris is 1.6248 metres tall.

b It took me 17 minutes 48.78 seconds to cook the dinner.

c My rabbit weighs 2.867 kg.

d The temperature at the bottom of the ocean is 1.239 °C.

e There were 23 736 people at the game yesterday.

2 How many jars each holding 119 cm^3 of water can be filled from a 3 litre flask?

3 If I walk at an average speed of 62 metres per minute, how long will it take me to walk a distance of 4 km?

4 Helen earns £31 500 a year. How much does she earn in

a 1 month **b** 1 week **c** 1 day?

5 Dave travelled a distance of 350 miles in 5 hours 40 minutes. What was his average speed?

6 Ten grams of Gold cost £2.17. How much will one kilogram of Gold cost?

10 Ratios, fractions, speed & proportion

HOMEWORK 10A

Example 1 Simplify 5 : 2 0.

5 : 20 = 1 : 4 (Divide both side of the ratio by 5.)

Example 2 Simplify 20p : £2.

(Change to a common unit) 20p : 200p = 1 : 10

Example 3 A garden is divided into lawn and shrubs in the ratio 3 : 2.

The lawn covers $\frac{3}{5}$ of the garden and the shrubs cover $\frac{2}{5}$ of the garden.

1 Express each of the following ratios in their simplest form.

a 3 : 9 **b** 5 : 25 **c** 4 : 24 **d** 10 : 30 **e** 6 : 9
f 12 : 20 **g** 25 : 40 **h** 30 : 4 **i** 14 : 35 **j** 125 : 50

2 Express each of the following ratios of quantities in their simplest form. (Remember to change to a common unit where necessary.)

a £2 to £8 **b** £12 to £16 **c** 25 g to 200 g
d 6 miles : 15 miles **e** 20 cm : 50 cm **f** 80p : £1.50
g 1 kg : 300g **h** 40 seconds : 2 minutes **i** 9 hours : 1 day
j 4 mm : 2 cm

3 £20 is shared out between Bob and Kathryn in the ratio 1 : 3.

a What fraction of the £20 does Bob receive?
b What fraction of the £20 does Kathryn receive?

4 In a class of students, the ratio of boys to girls is 2 : 3.

a What fraction of the class is boys?
b What fraction of the class is girls?

5 Pewter is an alloy containing lead and tin in the ratio 1 : 9.

a What fraction of pewter is lead?
b What fraction of pewter is tin?

HOMEWORK 10B

Example Divide £40 between Peter and Hitan in the ratio 2 : 3.

Changing the ratio to fractions gives
Peter's share = $\frac{2}{5}$ and Hitan's share = $\frac{3}{5}$
So, Peter receives $\frac{2}{5} \times$ £40 = £16 and Hitan receives $\frac{3}{5} \times$ £40 = £24.

1 Divide each of the following amounts in the given ratios.

a £10 in the ratio 1 : 4
b £12 in the ratio 1 : 2
c £40 in the ratio 1 : 3
d 60 g in the ratio 1 : 5
e 10 hours in the ratio 1 : 9
f 25 kg in the ratio 2 : 3
g 30 days in the ratio 3 : 2
h 70 m in the ratio 3 : 4
i £5 in the ratio 3 : 7
j 1 day in the ratio 5 : 3

2 The ratio of female to male members of a sports centre is 3 : 1. The total number of members of the centre is 400.

a How many members are female?
b How many members are male?

3 A 20 metre length of cloth is cut into two pieces in the ratio 1 : 9. How long is each piece?

4 James collects beer mats and the ratio of British mats to foreign mats is 5 : 2. He has 1400 beer mats in his collection. How many foreign beer mats does he have?

5 Patrick and Jane share out a box of sweets in the ratio of their ages. Patrick is 9 years old and Jane is 11 years old. If there are 100 sweets in the box, how many does Patrick get?

★**6** For her birthday Reena is given £30. She decides to spend four times as much as she saves. How much does she save?

7 Mrs Megson calculates that her quarterly electric and gas bills are in the ratio 5 : 6. The total she pays for both bills is £66. How much is each bill?

8 You can simplify a ratio by changing it into the form 1 : n. For example, 5 : 7 can be rewritten as 5 : 7 = 1 : 1.4 by dividing each side of the ratio by 5. Rewrite each of the following ratios in the form 1 : n.

a 2 : 3 **b** 2 : 5 **c** 4 : 5 **d** 5 : 8 **e** 10 : 21

HOMEWORK 10C

Example Two business partners, John and Ben, divided their total profit in the ratio 3 : 5. John received £2100. How much did Ben get?

John's £2100 was $\frac{3}{8}$ of the total profit.
So, $\frac{1}{8}$ of the total profit = £2100 ÷ 3 = £700.
Therefore, Ben's share, which was $\frac{5}{8}$, amounted to £700 × 5 = £3500.

1 Peter and Margaret's ages are in the ratio 4 : 5. If Peter is 16 years old, how old is Margaret?

2 Cans of lemonade and packets of crisps were bought for the school disco in the ratio 3 : 2. The organiser bought 120 cans of lemonade. How many packets of crisps did she buy?

3 In his restaurant, Manuel is making 'Sangria', a drink made from red wine and iced soda water, mixed in the ratio 2 : 3. Manuel uses 10 litres of red wine.

a How many litres of soda water does he use?
b How many litres of Sangria does he make?

4 Cupro-nickel coins are minted by mixing copper and nickel in the ratio 4 : 1.

a How much copper is needed to mix with 20 kg of nickel?
b How much nickel is needed to mix with 20 kg of copper?

5 The ratio of male to female spectators at a school inter-form football match is 2 : 1. If 60 boys watched the game, how many spectators were there in total?

★**6** Marmalade is made from sugar and oranges in the ratio 3 : 5. A jar of 'Savilles' marmalade contains 120 g of sugar.

a How many grams of oranges are in the jar?

b How many grams of marmalade are in the jar?

★**7** Each year Abbey School holds a sponsored walk for charity. The money raised is shared between a local charity and a national charity in the ratio 1 : 2. Last year the school gave £2000 to the local charity.

a How much did the school give to the national charity?

b How much did the school raise in total?

HOMEWORK 10D

The relationship between speed, time and distance can be expressed in three ways.

$$\text{Distance} = \text{Speed} \times \text{Time} \qquad \text{Speed} = \frac{\text{Distance}}{\text{Time}} \qquad \text{Time} = \frac{\text{Distance}}{\text{Speed}}$$

Example Sean is going to drive from Newcastle upon Tyne to Nottingham, a distance of 190 miles. He estimates that he will drive at an average speed of 50 mph. How long will it take him?

Sean's time = $\frac{190}{50}$ = 3.8 hours

Change the 0.8 hours to minutes by multiplying by 60, to give 48 minutes. So, the time for Sean's journey will be 3 hours 48 minutes.

Remember When you calculate a time and get a decimal answer, do not mistake the decimal part for minutes. You must either:

- leave the time as a decimal number and give the unit as hours, or
- change the decimal part to minutes by multiplying it by 60 (1 hour = 60 minutes) and give the answer in hours and minutes.

1 A cyclist travels a distance of 60 miles in 4 hours. What was her average speed?

2 How far along a motorway will you travel if you drive at an average speed of 60 mph for 3 hours?

3 Mr Baylis drives on a business trip from Manchester to London in $4\frac{1}{2}$ hours. The distance he travels is 207 miles. What is his average speed?

4 The distance from Leeds to Birmingham is 125 miles. The train I catch travels at an average speed of 50 mph. If I catch the 11.30am train in Leeds, at what time would I expect to be in Birmingham?

5 Copy and complete the following table.

	Distance travelled	Time taken	Average speed
a	240 miles	8 hours	
b	150 km	3 hours	
c		4 hours	5 mph
d		$2\frac{1}{2}$ hours	20 km/h
e	1300 miles		400 mph
f	90 km		25 km/h

★**6** A coach travels at an average speed of 60 km/h for 2 hours on a motorway and then slows down in a town centre to do the last 30 minutes of a journey at an average speed of 20 km/h.

a What is the total distance of this journey?

b What is the average speed of the coach over the whole journey?

★**7** Hilary cycles to work each day. She cycles the first 5 miles at an average speed of 15 mph and then cycles the last mile in 10 minutes.

a How long does it take Jane to get to work?

b What is her average speed for the whole journey?

★**8** Martha drives home from work in 1 hour 15 minutes. She drives home at an average speed of 36 mph.

a Change 1 hour 15 minutes to decimal time in hours.

b How far is it from Martha's work to her home?

HOMEWORK 10E

Example If eight pens cost £2.64, what is the cost of five pens?

First find the cost of one pen. This is £2.64 ÷ 8 = £0.33.
The cost of five pens is £0.33 × 5 = £1.65.

1 If four video tapes cost £3.20, what would ten video tapes cost?

2 Five oranges cost 90p. Find the cost of twelve oranges.

3 Dylan earns £18.60 in 3 hours. How much will he earn in 8 hours?

4 Barbara bought 12 postcards for 3 euros when she was on holiday in Tenerife.

a How many euros would she have paid if she had only bought 9 postcards?

b How many postcards could she have bought with a 5 euro note?

5 Five 'Day-Rover' bus tickets cost £8.50.

a What is the cost of 16 tickets?

b How many tickets can be bought for £20?

6 A car uses 8 litres of petrol on a trip of 72 miles.

a How much would be used on a trip of 54 miles?

b How far would the car go on a full tank of 45 litres?

7 It takes a photocopier 18 seconds to produce 12 copies. How long will it take to produce 32 copies?

★**8** Val has a recipe for making 12 flapjacks.

100 g margarine
4 tablespoons golden syrup
80 g granulated sugar
200 g rolled oats

a What is the recipe for:

i 6 flapjacks **ii** 24 flapjacks **iii** 30 flapjacks?

b What is the maximum number of flapjacks she can make if she has 1 kg of each ingredient?

HOMEWORK 10F

Example There are two different-sized packets of Whito soap powder at a supermarket. The medium size contains 800 g and costs £1.60 and the large size contains 2.5 kg and costs £4.75. Which is the better buy?

Find the weight per unit cost for both packets.

Medium: $800 \div 160 = 5$ g per pence

Large: $2500 \div 475 = 5.26$ g per pence

From these we see that there is more weight per pence with the large size, which means that the large size is the better buy.

1 Compare the following pairs of product and state which is the better buy and why.

a Tomato ketchup: a medium bottle which is 200 g for 55p or a large bottle which is 350 g for 87p.

b Milk chocolate: a 125 g bar at 77p or a 200 g bar at 92p.

c Coffee: a 750 g tin at £11.95 or a 500 g tin at £7.85.

d Honey: a large jar which is 900 g for £2.35 or a small jar which is 225 g for 65p.

2 Boxes of 'Wetherels' teabags are sold in three different sizes.

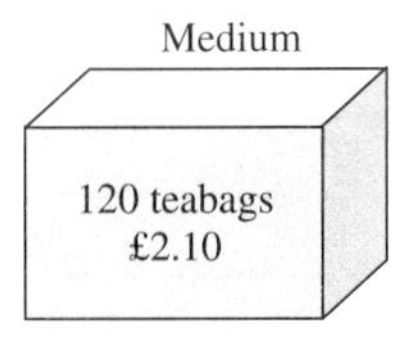

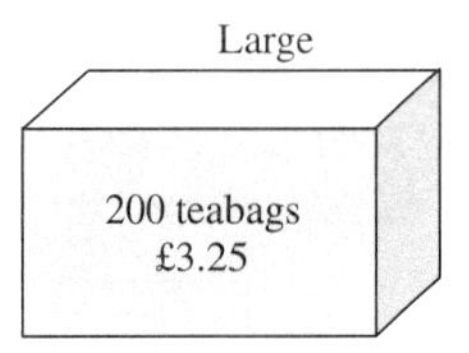

Which size of teabags gives the best value for money?

3 Bottles of 'Cola' are sold in different sizes. Copy and complete the table.

Size of bottle	Price	Cost per litre
$\frac{1}{2}$ litre	36p	
$1\frac{1}{2}$ litres	99p	
2 litres	£1.40	
3 litres	£1.95	

Which bottle gives the best value for money?

★**4** The following 'special offers' were being promoted by a supermarket.

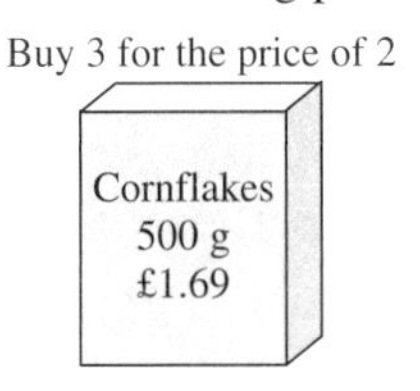

Which offer is the better value for money? Explain why.

11 Percentages

HOMEWORK 11A

Example 1 As a fraction 32% = $\frac{32}{100}$ which can be cancelled down to $\frac{8}{25}$

Example 2 As a decimal 65% = 65 ÷ 100 = 0.65

1 Write each percentage as a fraction in its lowest terms.

a 10% **b** 40% **c** 25% **d** 15% **e** 75% **f** 35%
g 12% **h** 28% **i** 56% **j** 18% **k** 42% **l** 6%

2 Write each percentage as a decimal.

a 87% **b** 25% **c** 33% **d** 5% **e** 1% **f** 72%
g 58% **h** 17.5% **i** 8.5% **j** 68.2% **k** 150% **l** 132%

3 Copy and complete the table.

Percentage	Fraction	Decimal
10%		
20%		
30%		
		0.4
		0.5
		0.6
	$\frac{7}{10}$	
	$\frac{8}{10}$	
	$\frac{9}{10}$	

4 If 45% of pupils walk to school, what percentage do not walk to school?

5 If 84% of the families in a village own at least one car, what percentage of the families do not own a car?

6 In a local election, of all the people who voted, 48% voted for Mrs Slater, 29% voted for Mr Rhodes and the remainder voted for Mr Mulley. What percentage voted for Mr Mulley?

7 From his gross salary, Mr Hardy pays 20% Income Tax, 6% Superannuation and 5% National Insurance. What percentage is his net pay?

8 Approximately what percentage of each can is filled with oil?

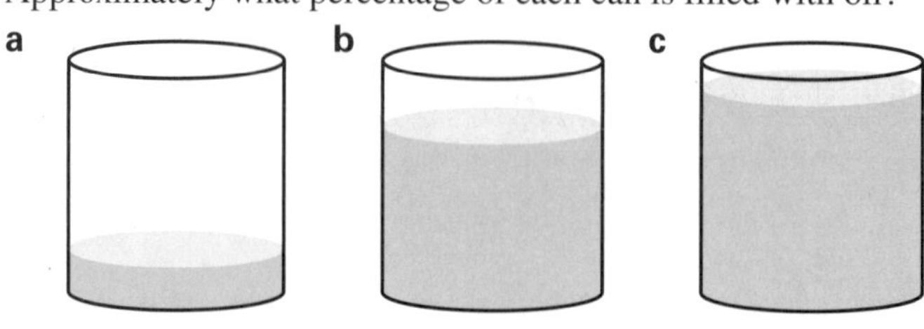

9 Write each fraction as a percentage.

a $\frac{3}{4}$ **b** $\frac{2}{5}$ **c** $\frac{7}{20}$ **d** $\frac{3}{25}$ **e** $\frac{43}{50}$ **f** $\frac{3}{8}$

10 Write each decimal as a percentage.

a 0.23 **b** 0.87 **c** 0.09 **d** 0.235 **e** 1.8 **f** 2.34

11 Tom scored 68 marks out of a possible 80 marks in a Geography test.

a Write his score as a fraction in its simplest form.

b Write his score as a decimal.

c Write his score as a percentage.

HOMEWORK 11B

Example Calculate 12% of 54 kg.

Method 1 $12 \div 100 \times 54 = 6.48$ kg

Method 2 Using a multiplier: $0.12 \times 54 = 6.48$ kg

1 What multiplier is equivalent to a percentage of:

a 23% **b** 70% **c** 4% **d** 120%?

2 What percentage is equivalent to a multiplier of:

a 0.38 **b** 0.8 **c** 0.07 **d** 1.5?

3 Calculate the following.

a 25% of £200 **b** 10% of £120 **c** 53% of 400 kg **d** 75% of 84 cm

e 22% of £84 **f** 71% of 250 g **g** 24% of £3 **h** 95% of 320 m

i 6% of £42 **j** 17.5% of £56 **k** 8.5% of 160 *l* **l** 37.2% of £800

4 During one week at a Test Centre, 320 people took their driving test and 65% passed. How many people passed?

5 A school has 250 pupils on roll in each year and the attendance record on one day for each year group is shown below.

Year 7 96%, Year 8 92%, Year 9 84%, Year 10 88%, Year 11 80%

How many pupils were present in each year group on that day?

6 A certain type of stainless steel consists of 84% iron, 14% chromium and 2% carbon (by weight). How much of each is in 450 tonnes of stainless steel?

★**7** Value Added Tax (VAT) is added on to most goods purchased at the rate of 17.5%. How much VAT will be added on to the following bills:

a a restaurant bill for £40 **b** a telephone bill for £82

c a car repair bill for £240?

★**8** An insurance firm sells house insurance and the annual premiums are usually at a cost of 0.5% of the value of the house. What will be the annual premium for a house valued at £120 000?

HOMEWORK 11C

Example Increase £6 by 5%.

Method 1 Find 5% of £6: $(5 \div 100) \times 6 = £0.30$

Add the £0.30 to the original amount: $£6 + £0.30 = £6.30$

Method 2 Using a multiplier: $1.05 \times 6 = £6.30$

1 Increase each of the following by the given percentage. (Use any method you like.)

a	£80 by 5%	**b**	£150 by 10%	**c**	800 m by 15%	**d**	320 kg by 25%
e	£42 by 30%	**f**	£24 by 65%	**g**	120 cm by 18%	**h**	£32 by 46%
i	550g by 85%	**j**	£72 by 72%				

2 Mr Kent, who was on a salary of £32 500, was given a pay rise of 4%. What is his new salary?

3 Copy and complete this electricity bill.

	Total charges
Fixed charges	£13.00
840 units @ 6.45 p per unit	
1720 units @ 2.45 p per unit	
Total charges	
VAT @ 8%	
Total to pay	

4 A bank pays 8% simple interest on the money that each saver keeps in a savings account for a year. Miss Pettica puts £2000 in this account for three years. How much will she have in her account after:

a 1 year **b** 2 years **c** 3 years?

★**5** VAT is a tax that the Government adds to the price of goods sold. At the moment it is 17.5% on all goods. Calculate the price of the following gifts Mrs Dow purchased from a gift catalogue, after VAT of 17.5% has been added.

Gift	*Pre-VAT price*
Travel alarm clock	£18.00
Ladies' purse wallet	£15.20
Pet's luxury towel	£12.80
Silver-plated bookmark	£6.40

HOMEWORK 11D

Example Decrease £6 by 5%.

Method 1 Find 5% of £6: $(5 \div 100) \times 6 = £0.30$
Subtract the £0.30 from the original amount: $£6 - £0.30 = £5.70$

Method 2 Using a multiplier: $0.95 \times 6 = £5.70$

1 Decrease each of the following by the given percentage. (Use any method you like.)

a	£20 by 10%	**b**	£150 by 20%	**c**	90 kg by 30%	**d**	500 m by 12%
e	£260 by 5%	**f**	80 cm by 25%	**g**	400 g by 42%	**h**	£425 by 23%
i	48 kg by 75%	**j**	£63 by 37%				

2 Mrs Denghali buys a new car from a garage for £8400. The garage owner tells her that the value of the car will lose 24% after one year. What will be the value of the car after one year?

★**3** The population of a village in 2001 was 2400. In 2006 the population had decreased by 12%. What was the population of the village in 2006?

★**4** A Travel Agent is offering a 15% discount on holidays. How much will the advertised holiday now cost?

NEW YORK FOR A WEEK
£540

★**5** **New Year's Sale:**
All prices reduced by 20%

Find the sale price of the following goods in the sale.

a a shirt at £30 **b** a suit at £130 **c** a pair of shoes at £42

HOMEWORK 11E

Example Express £6 as a percentage of £40.

Set up the fraction $\frac{6}{40}$ and multiply it by 100. $6 \div 40 = 15\%$.

1 Express each of the following as a percentage. Give your answers to one decimal place where necessary.

a £8 of £40 **b** 20 kg of 80 kg **c** 5 m of 50 m
d £15 of £20 **e** 400 g of 500 g **f** 23 cm of 50 cm
g £12 of £36 **h** 18 minutes of 1 hour **i** £27 of £40
j 5 days of 3 weeks

2 What percentage of these shapes is shaded?

a 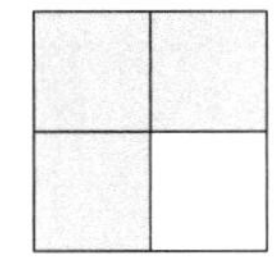 **b** 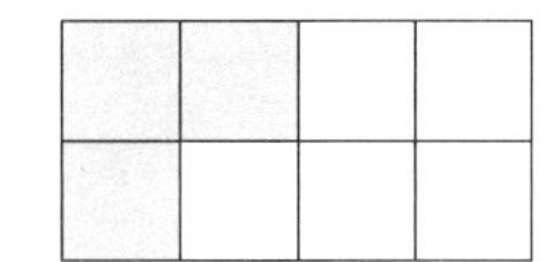

3 In a class of 30 pupils, 18 are girls.

a What percentage of the class are girls?
b What percentage of the class are boys?

4 The area of a farm is 820 hectares. The farmer uses 240 hectares for pasture. What percentage of the farm land is used for pasture? Give your answer to one decimal place.

5 Find, to one decimal place, the percentage profit on each of the following.

	Item	*Retail price (Selling price)*	*Wholesale price (Price the shop paid)*
a	Micro Hi-Fi System	£250	£150
b	CD Radio Cassette	£90	£60
c	MiniDisc Player	£44.99	£30
d	Cordless Headphones	£29.99	£18

12 Quadratic graphs

HOMEWORK 12A

1 a Copy and complete the table for the graph of $y = 2x^2$ for $-3 \leqslant x \leqslant 3$.

x	–3	–2	–1	0	1	2	3
$y = 2x^2$	18		2			8	

b Use the graph to find the value of y when $x = -1.4$.

c Use the graph to find the values of x that give a y-value of 10.

2 a Copy and complete the table for the graph of $y = x^2 + 3$ for $-5 \leqslant x \leqslant 5$.

x	–5	–4	–3	–2	–1	0	1	2	3	4	5
$y = x^2 + 3$	28		12					7			28

b Use the graph to find the value of y when $x = 2.5$.

c Use the graph to find the values of x that give a y-value of 10.

★**3 a** Copy and complete the table for the graph of $y = x^2 - 3x + 2$ for $-3 \leqslant x \leqslant 4$.

x	–3	–2	–1	0	1	2	3	4
$y = x^2 - 3x + 2$	20			2			2	

b Use the graph to find the value of y when $x = -1.5$.

c Use the graph to find the values of x that give a y-value of 2.5.

HOMEWORK 12B

1 a Copy and complete the table to draw the graph of $y = x^2 - 3x + 2$ for $-1 \leqslant x \leqslant 5$.

x	–1	0	1	2	3	4	5
$y = x^2 - 3x + 2$	6	2			2		

b Use your graph to find the solutions of the equation $x^2 - 3x + 2 = 0$.

2 a Copy and complete the table to draw the graph of $y = x^2 - 5x + 4$ for $-1 \leqslant x \leqslant 6$.

x	–1	0	1	2	3	4	5	6
$y = x^2 - 5x + 4$	10	4				0		

b Use your graph to find the solutions of the equation $x^2 - 5x + 4 = 0$.

3 a Copy and complete the table to draw the graph of $y = x^2 + 4x - 6$ for $-5 \leqslant x \leqslant 2$.

x	–5	–4	–3	–2	–1	0	1	2
$y = x^2 + 4x - 6$	–1							6

b Use your graph to find the solutions of the equation $x^2 + 4x - 6 = 0$.

13 Perimeter and area

HOMEWORK 13A

1 Calculate the perimeter of each of the following shapes.

a

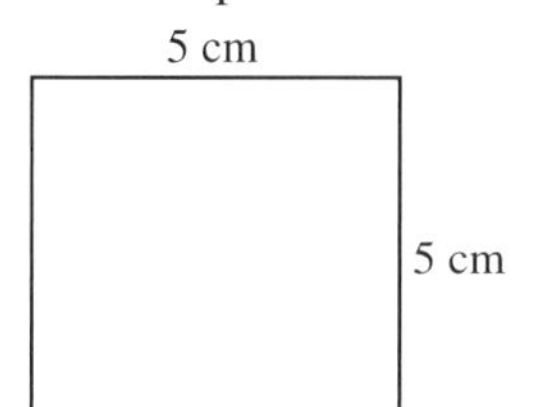

b

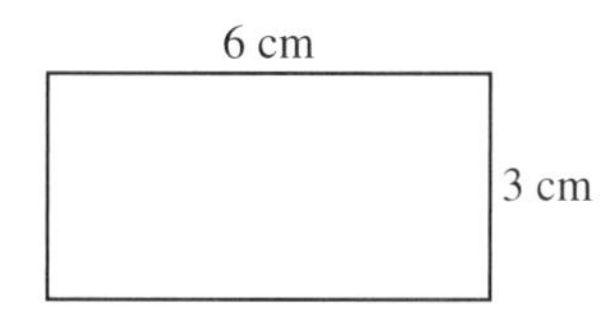

c

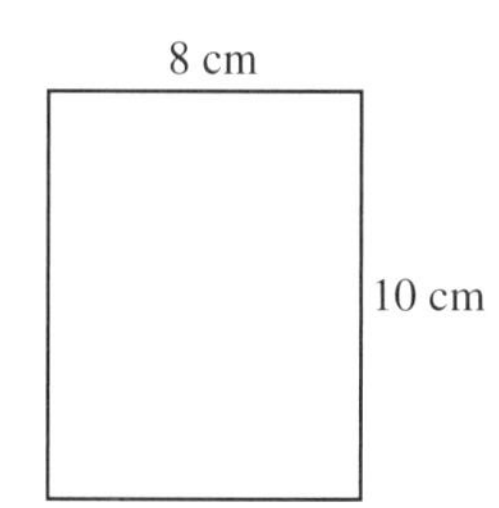

d

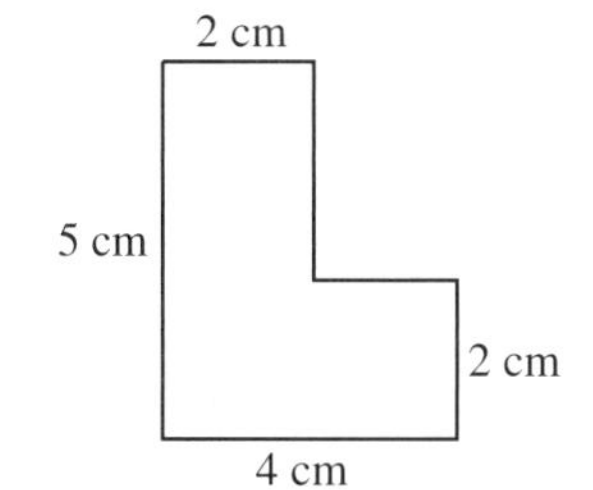

e

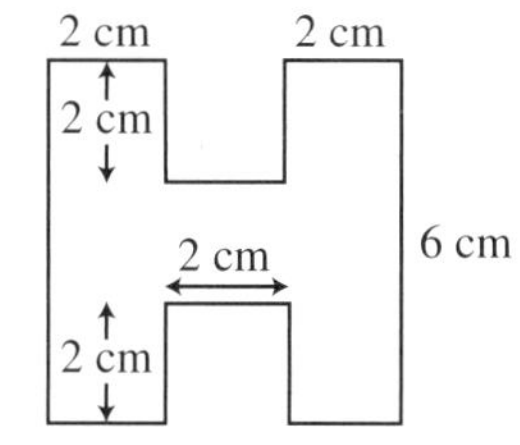

f

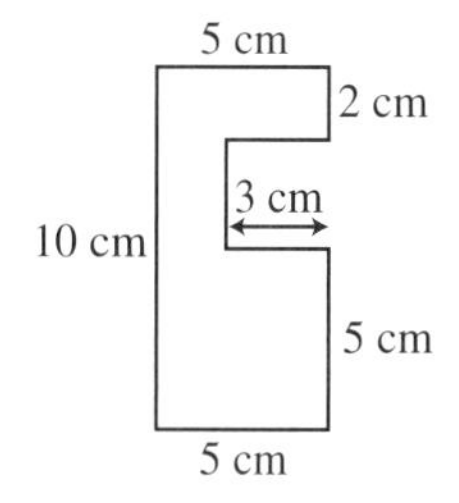

2 Draw as many different rectangles as possible with a perimeter of 14 cm.

3 Is it possible to draw a rectangle with a perimeter of 9 cm? Explain your answer.

HOMEWORK 13B

1 By counting squares, find the area of each of these shapes, giving answers in cm^2.

a

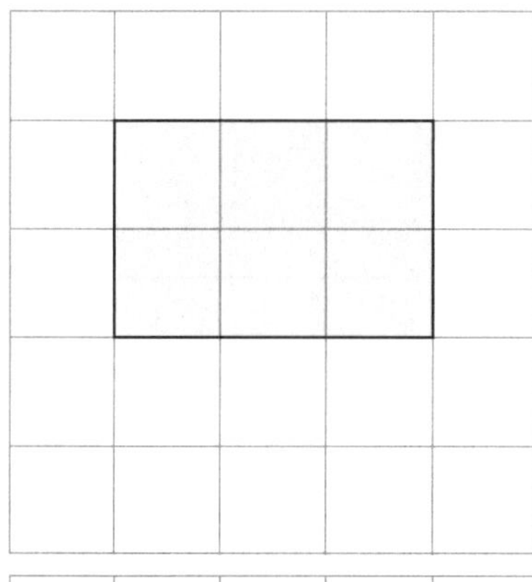

b

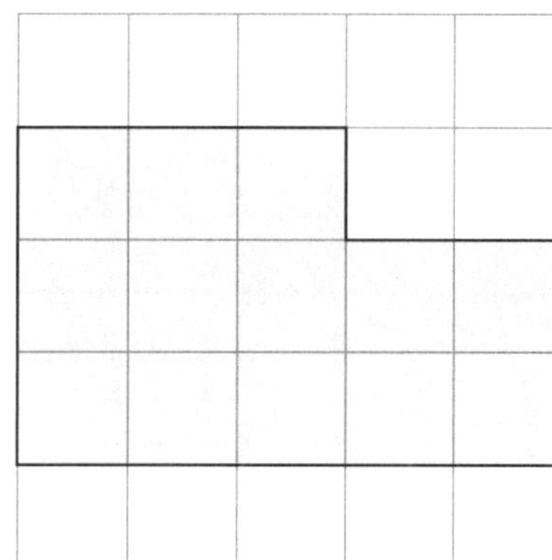

c

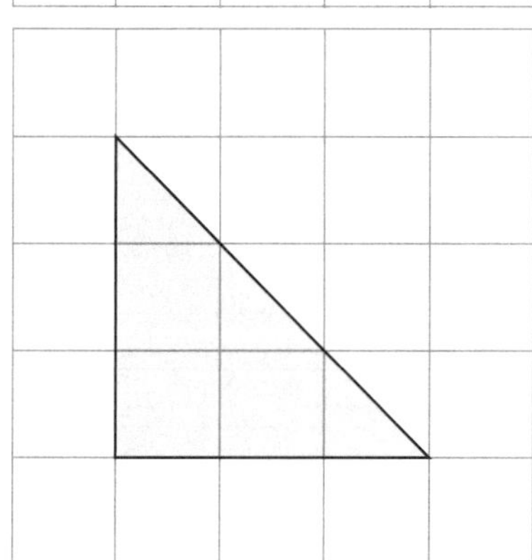

d

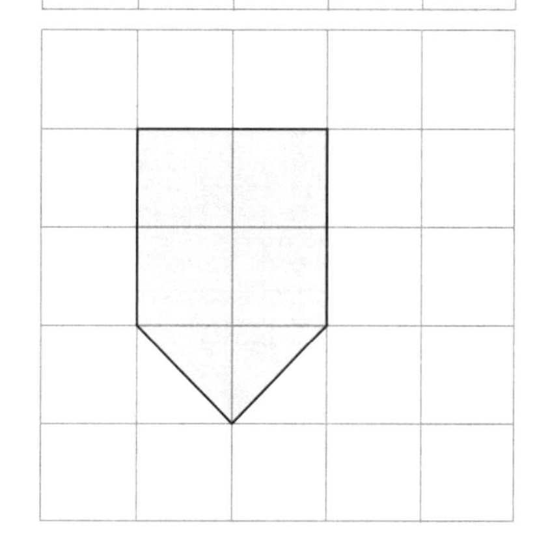

2 By counting squares, estimate the area of each of these shapes, giving answers in cm^2.

a

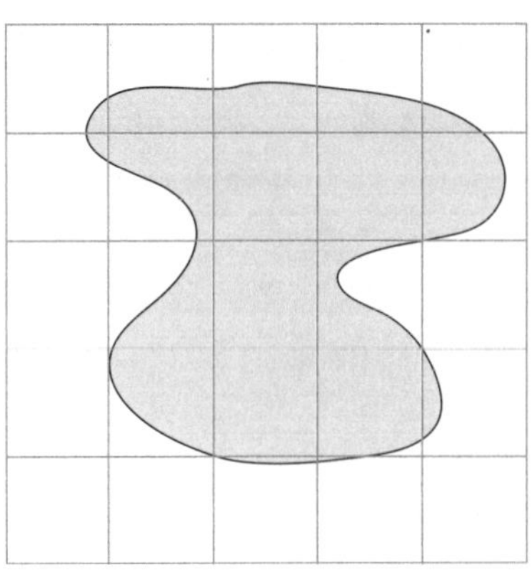

b

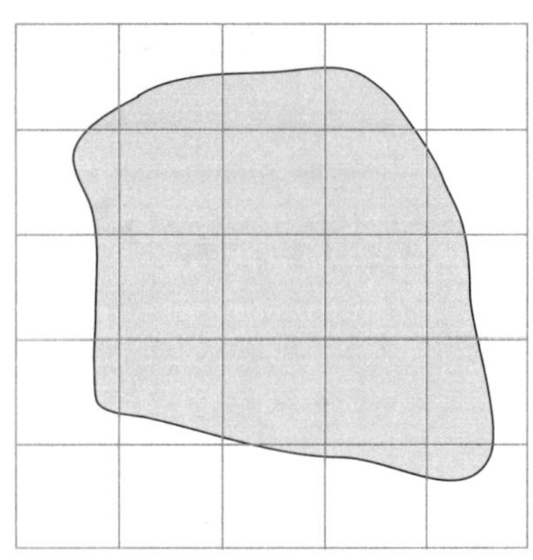

c

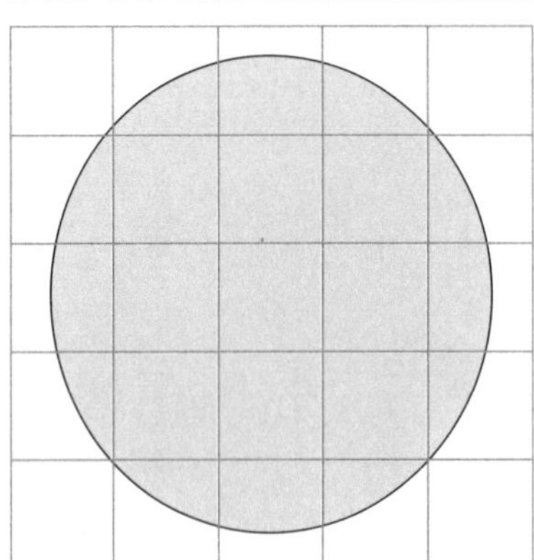

d

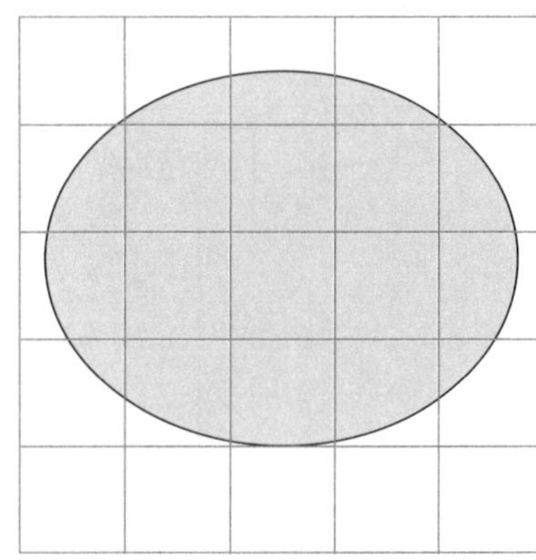

HOMEWORK 13C 

1 Calculate the area and the perimeter of each rectangle below.

a 5 cm, 2 cm

b

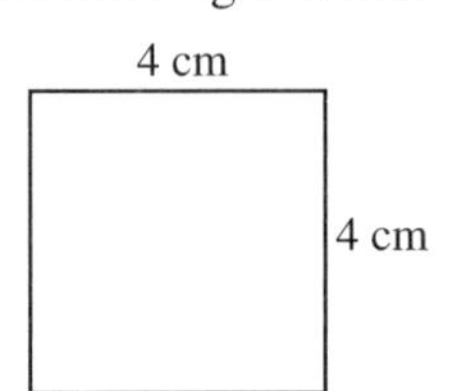

c

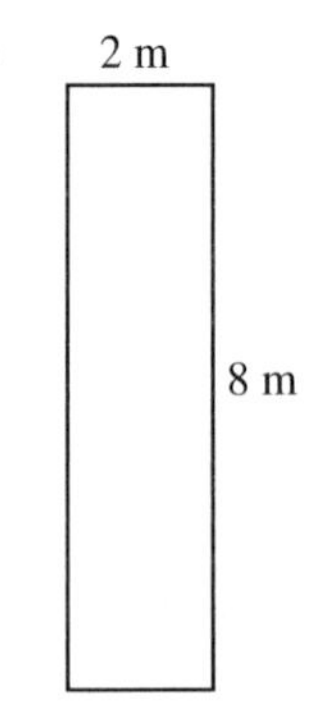

d 12 mm, 3 mm

e 20 m, 10 m

2 Copy and complete the following table for rectangles **a** to **e**.

	Length	Width	Perimeter	Area
a	4 cm	2 cm		
b	7 cm	4 cm		
c	6 cm		22 cm	
d		3 cm		15 cm^2
e			30 cm	50 cm^2

3 Copy and complete the statements below.

a 1 cm^2 = …… mm^2 **b** 1 m^2 = …… cm^2

HOMEWORK 13D

Calculate the area of each shape below as follows.

- First, split it into rectangles.
- Then, calculate the area of each rectangle.
- Finally, add together the areas of the rectangles.

1

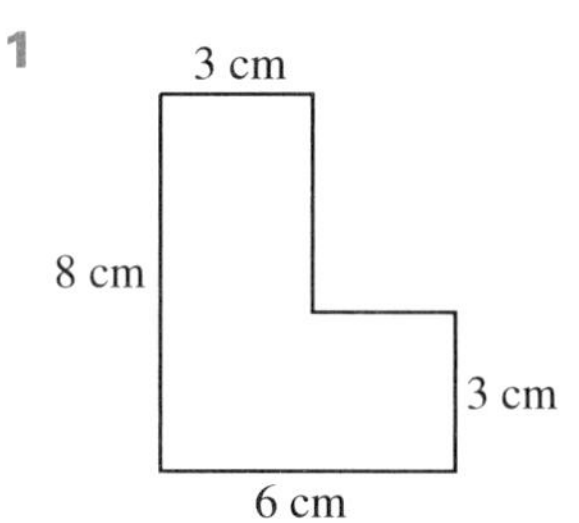

2

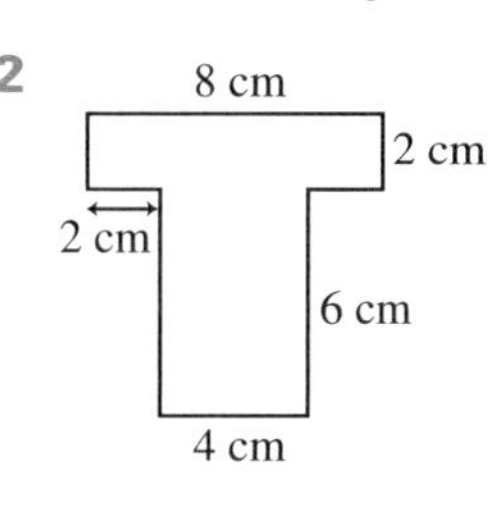

3

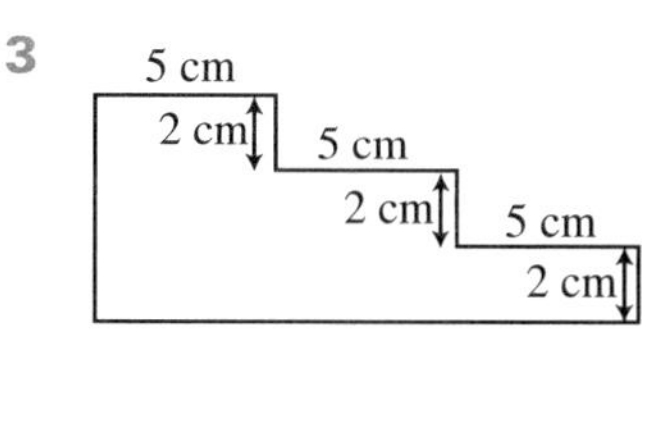

4

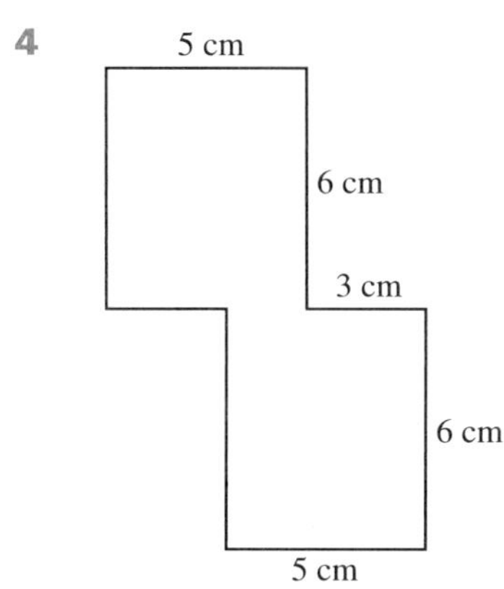

5

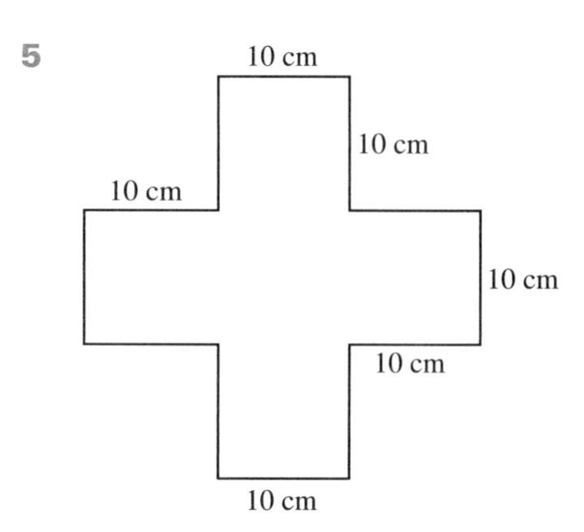

HOMEWORK 13E

Example Find the area of this triangle

$$\text{Area} = \tfrac{1}{2} \times 7 \times 4$$
$$= \tfrac{1}{2} \times 28 = 14 \text{ cm}^2$$

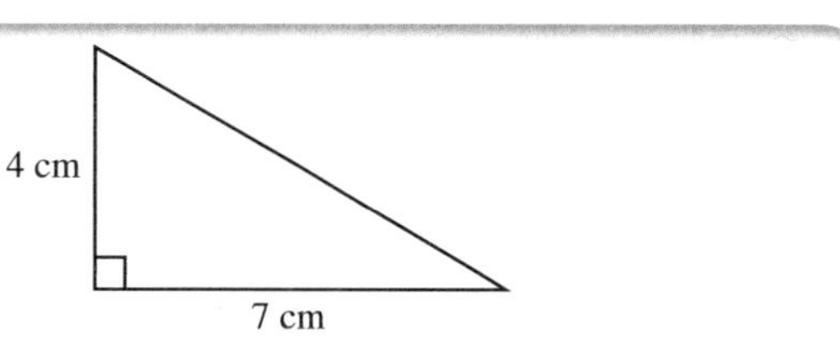

1 Write down the perimeter and area of each triangle.

a

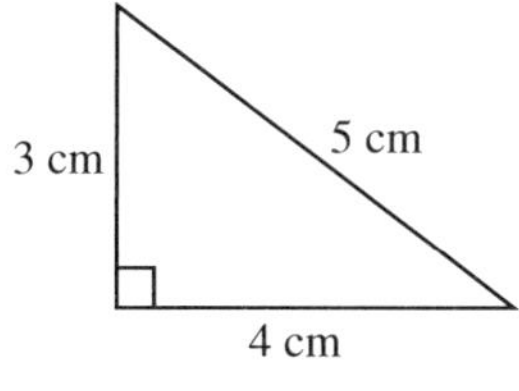

b

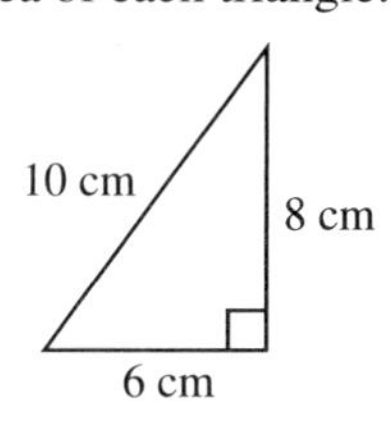

c

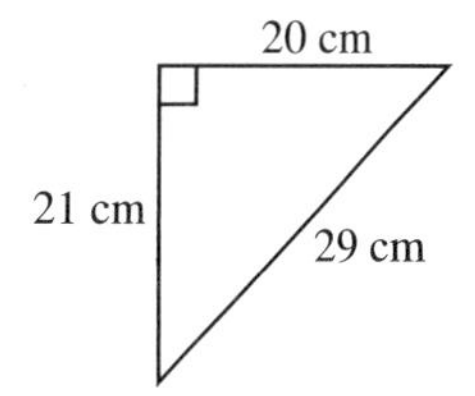

2 Find the areas of these composite shapes.

a

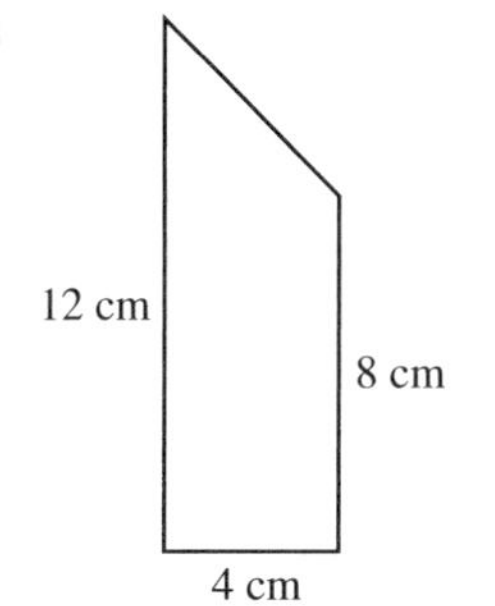

b

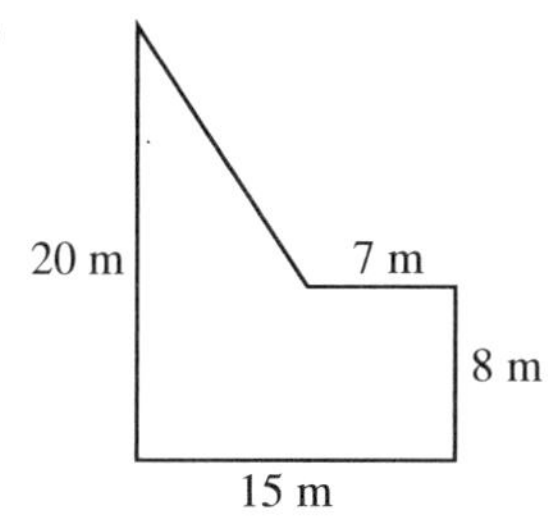

c

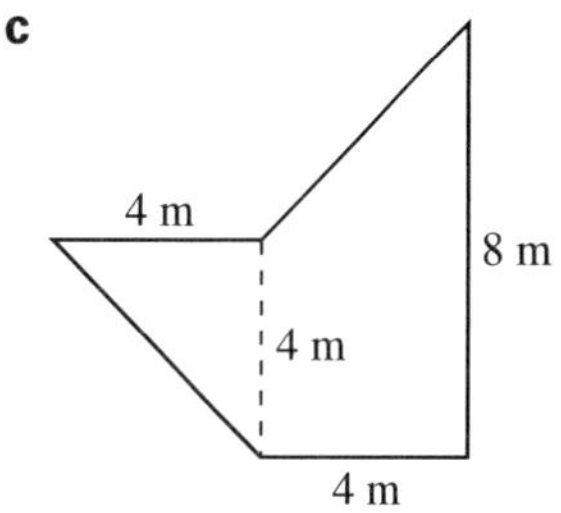

★**3** Find the area of the wood on this blackboard 90° set square.

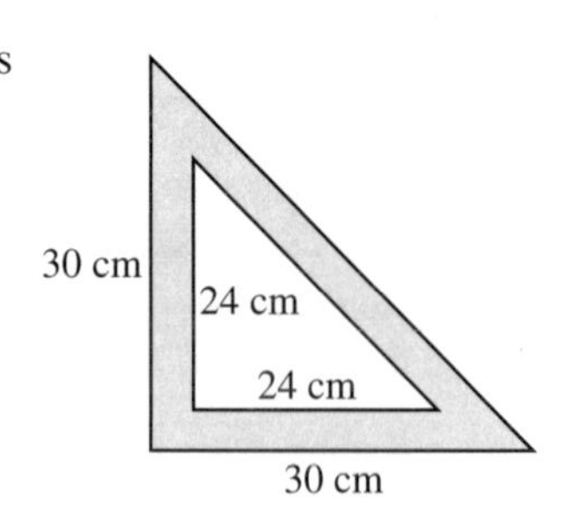

★**4** Which of these three triangles has the smallest area?

a

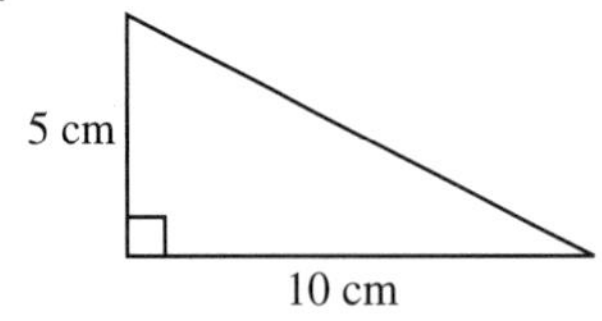

b

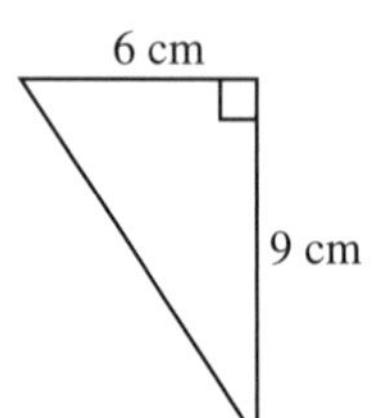

c

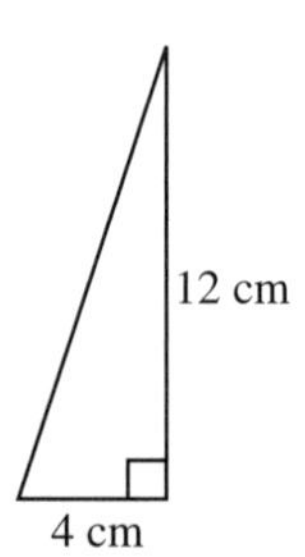

HOMEWORK 13F

Example Find the area of this triangle.

$$\text{Area} = \tfrac{1}{2} \times 9 \times 4$$
$$= \tfrac{1}{2} \times 36 = 18 \text{ cm}^2$$

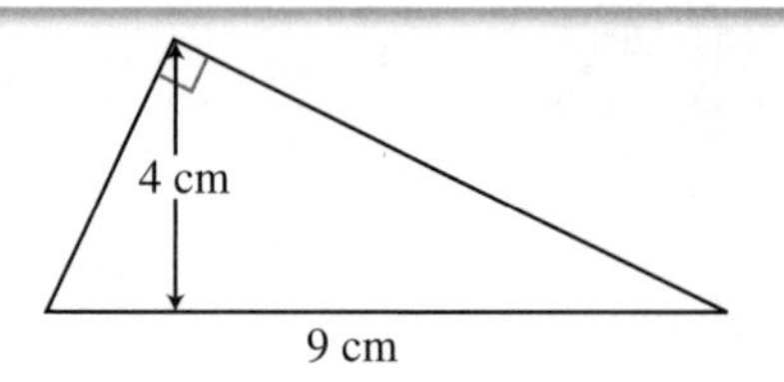

1 Calculate the area of each of these triangles.

a

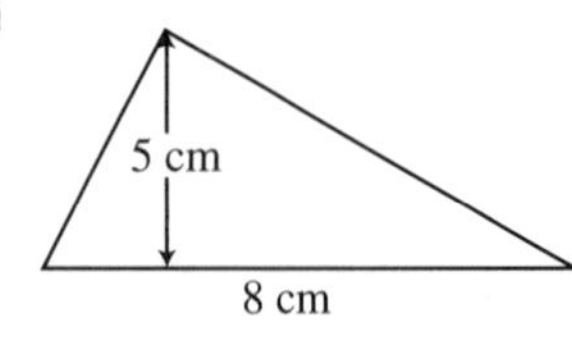

b

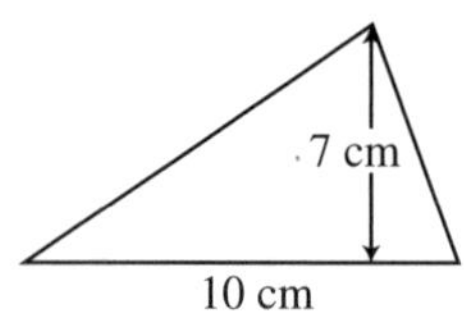

c

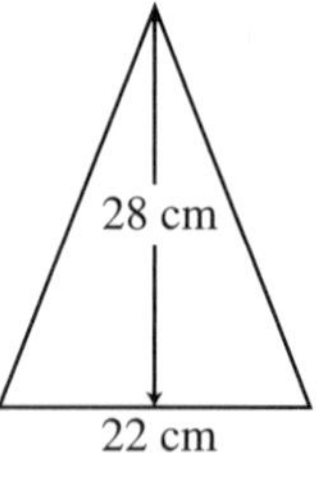

d

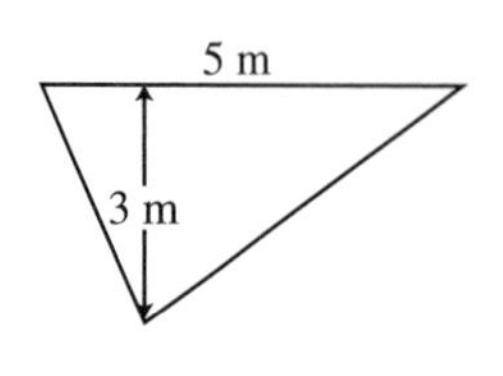

e

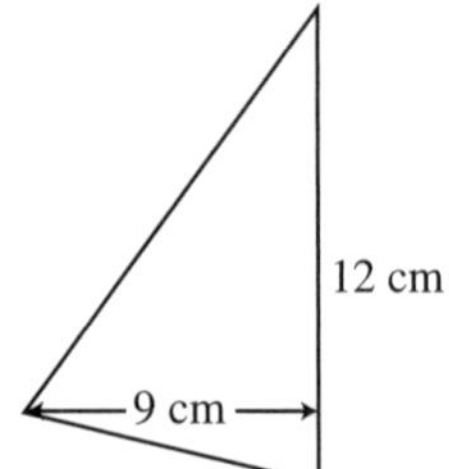

f

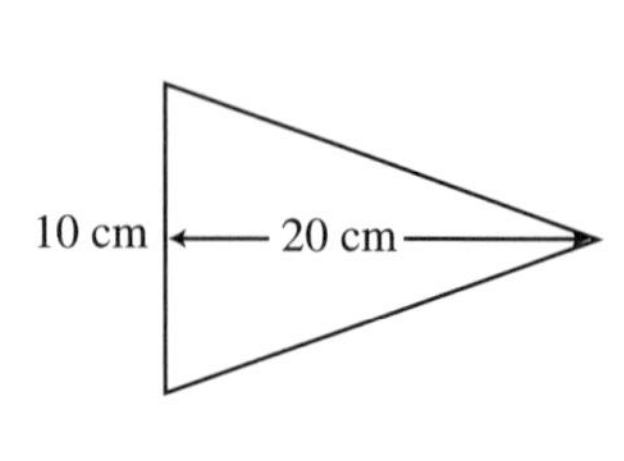

2 Copy and complete the following table for triangles **a** to **e**.

	Base	Vertical height	Area
a	6 cm	8 cm	
b	10 cm	7 cm	
c	5 cm	5 cm	
d	4 cm		12 cm^2
e		20 cm	50 cm^2

★**3** Find the area of each of the shaded shapes.

a 40 cm, 10 cm, 80 cm

b 6 cm, 6 cm, 8 cm, 8 cm

c 12 cm, 12 cm, 7 cm, 8 cm

★**4** Draw diagrams to show two different-sized triangles that have the same area of 40 cm^2.

HOMEWORK 13G

Example Find the area of this parallelogram.

$$\text{Area} = 8 \times 6$$
$$= 48 \text{ cm}^2$$

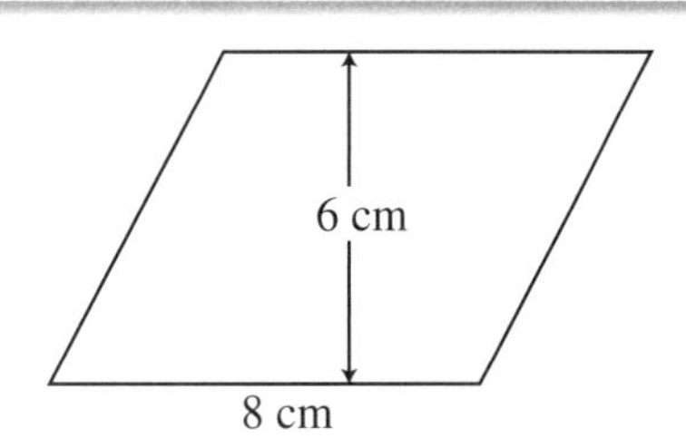

1 Calculate the area of each parallelogram below.

a

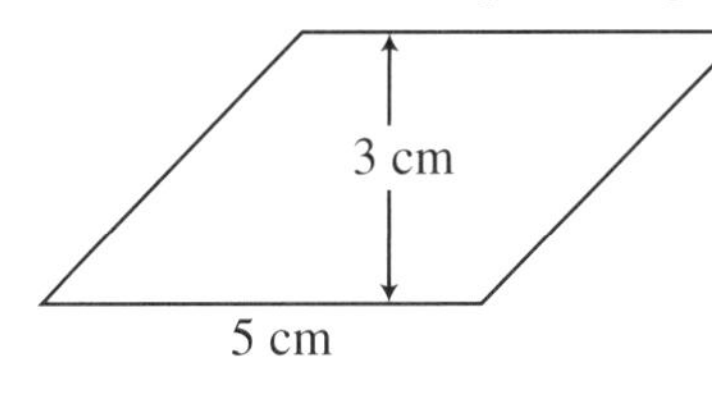

b

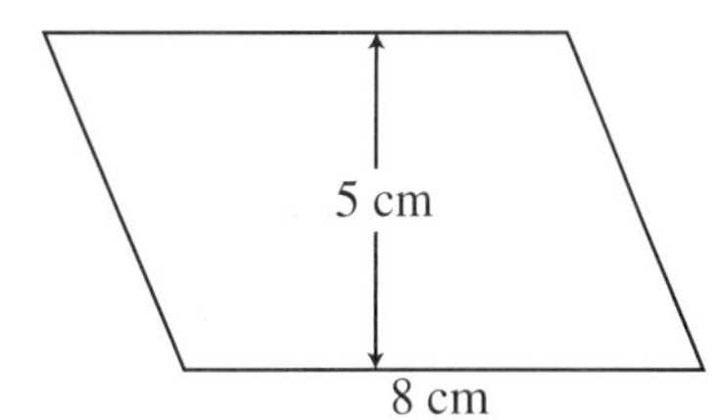

c

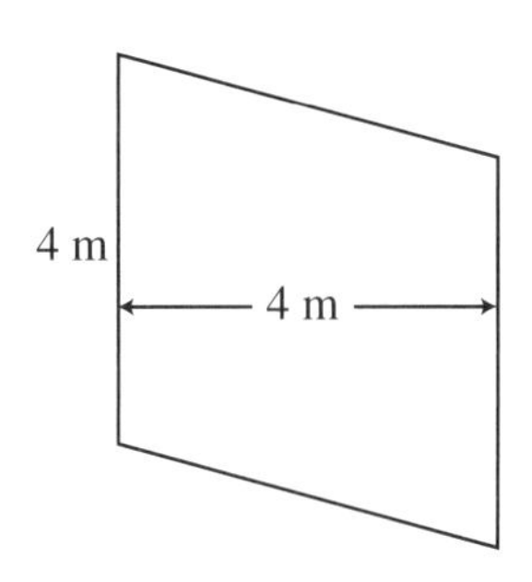

d

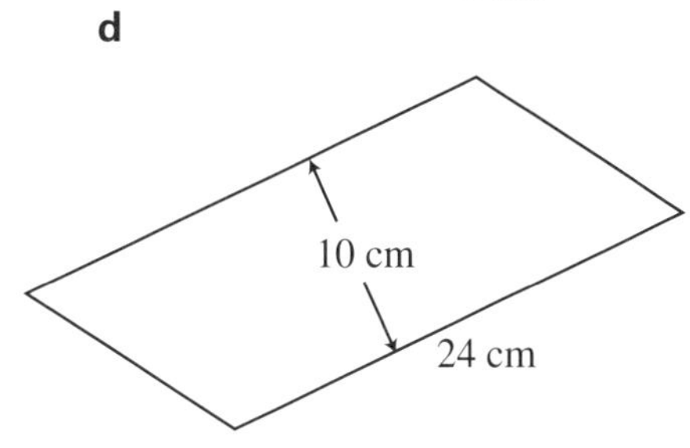

2 Find the area of the shaded shape.

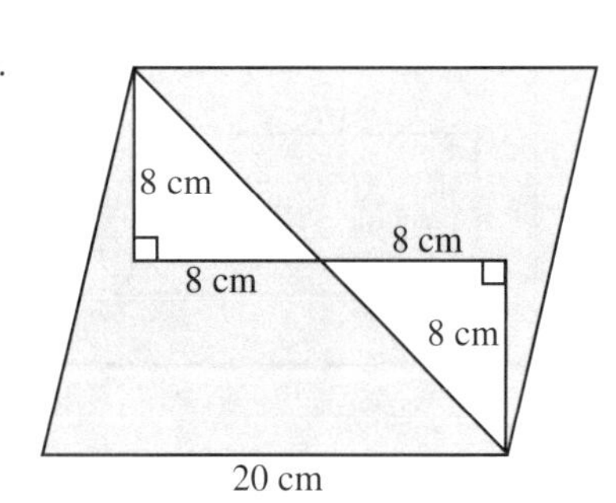

HOMEWORK 13H

1 Calculate the perimeter and the area of each of these trapeziums.

a

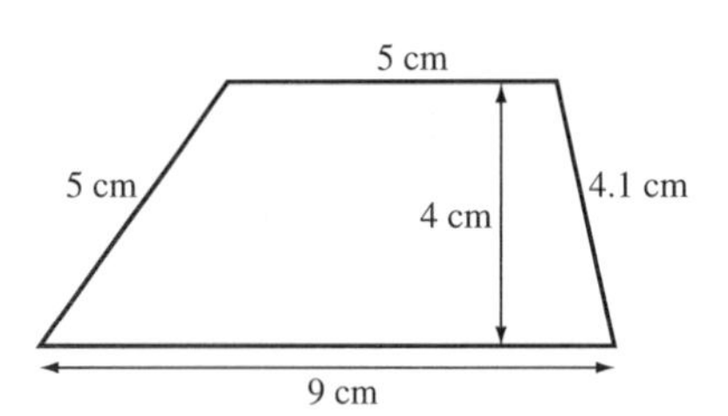

b

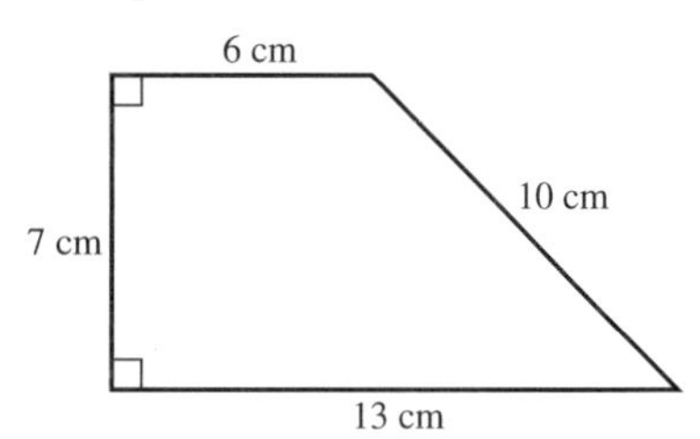

2 Calculate the area of each of these shapes.

a

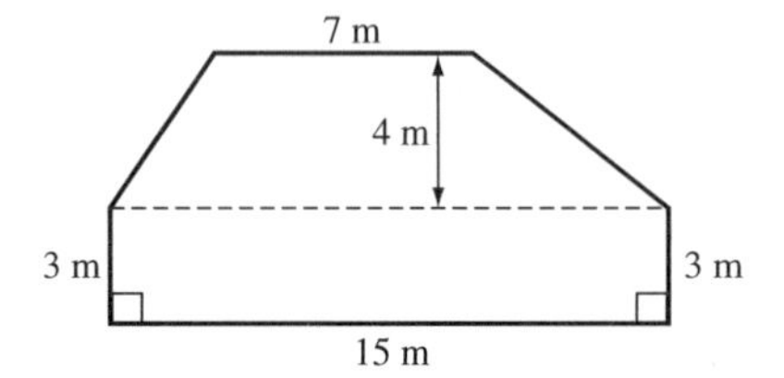

b

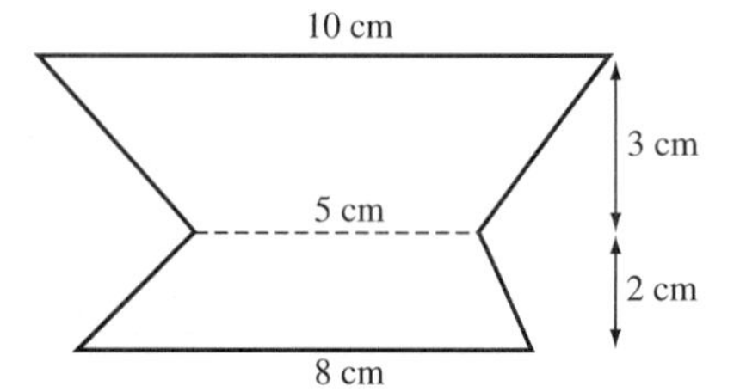

3 Calculate the area of the shaded part in each of these diagrams.

a

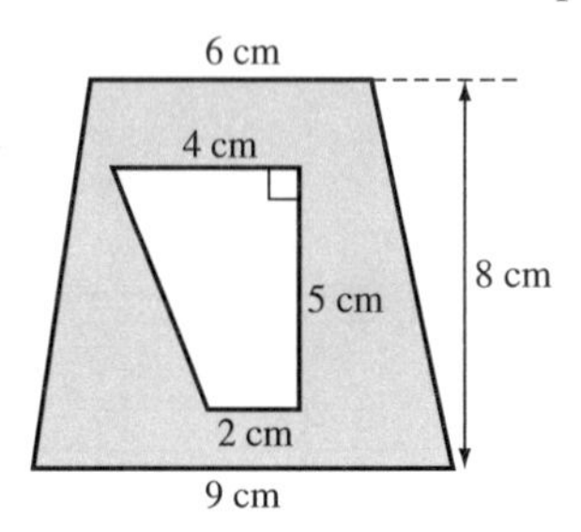

b

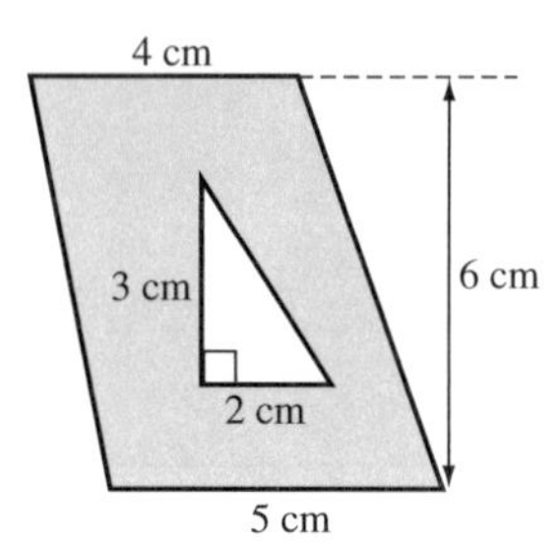

4 Which of the following shapes has the larger area?

a

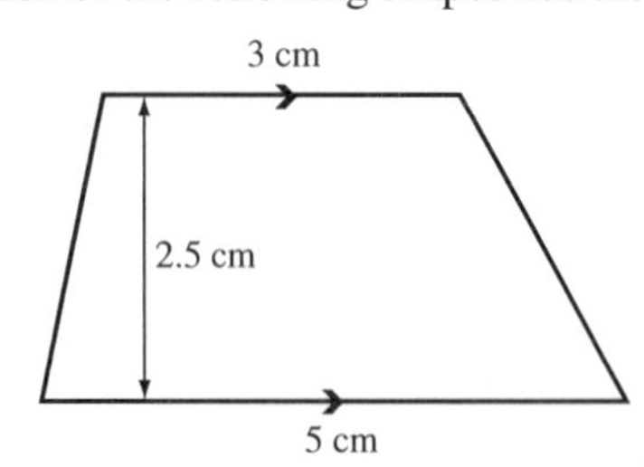

b

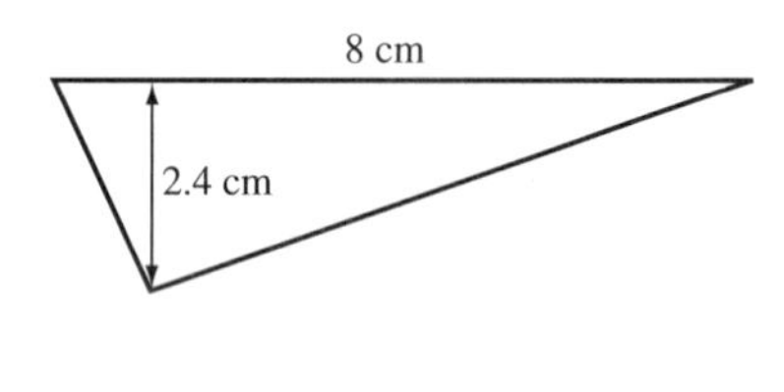

5 Calculate the area of the shaded part in the diagram.

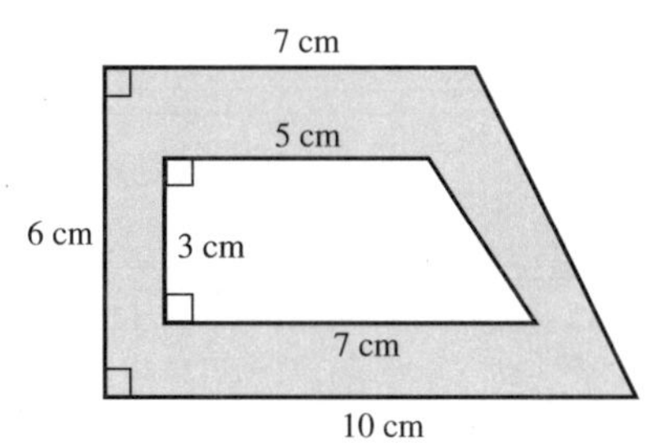

14 Symmetry

HOMEWORK 14A 

1 Copy these shapes and draw on the lines of symmetry for each one. If it will help you, use tracing paper or a mirror to check your answers.

a 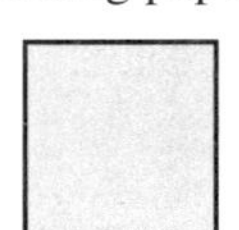 **b**  **c** 

d 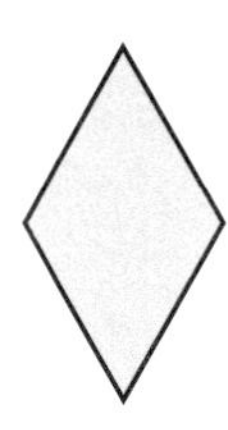 **e** 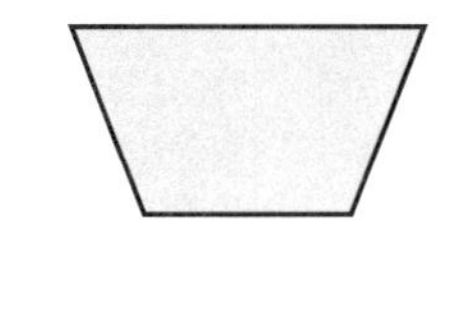

2 Copy this regular hexagon and draw on all the lines of symmetry.

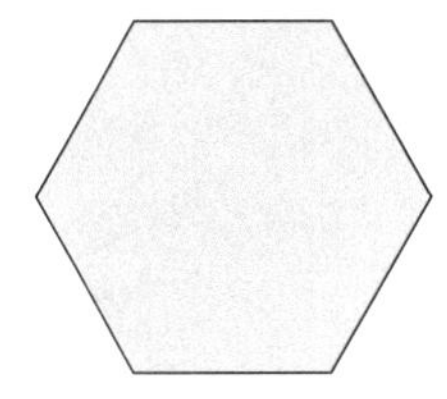

3 Copy these flow chart symbols and draw on all the lines of symmetry for each one.

a 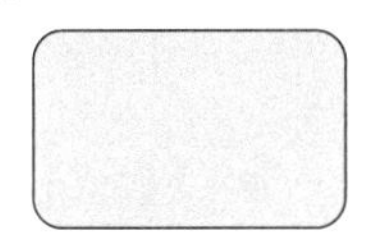 **b**  **c** 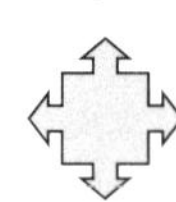

d 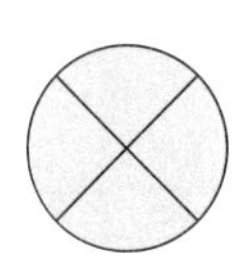 **e** 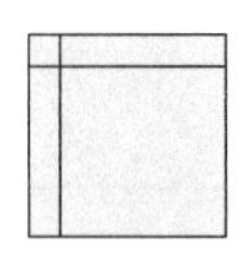

4 Write down the number of lines of symmetry for each of these flags.

a  **b**  **c** 

5 How many lines of symmetry do each of these letters have?

a  **b**  **c** H **d** T **e** Y

★6 Draw three copies of the diagram below.

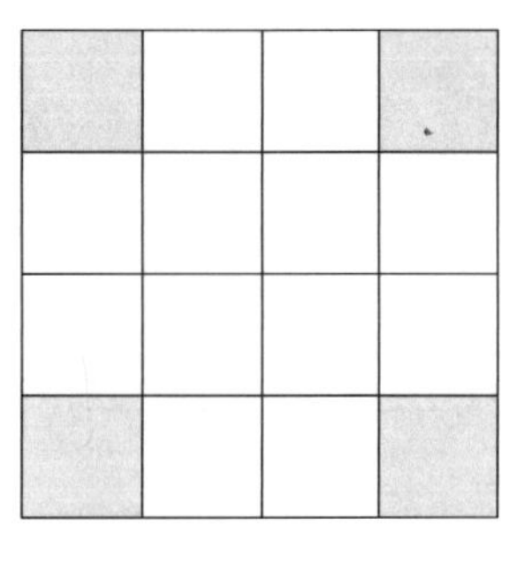

- **a** Shade in two more squares so that the diagram has no lines of symmetry.
- **b** Shade in two more squares so that the diagram has exactly one line of symmetry.
- **c** Shade in two more squares so that the diagram has exactly two lines of symmetry.

HOMEWORK 14B

1 Copy these shapes and write below each one the order of rotational symmetry. If it will help you, use tracing paper.

a 

b 

c 

d 

e 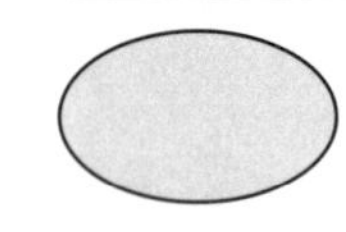

2 Write down the order of rotational symmetry for each of these shapes.

a 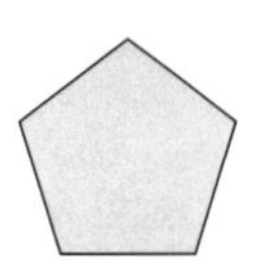

b 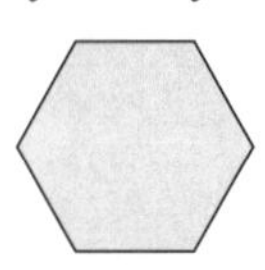

c 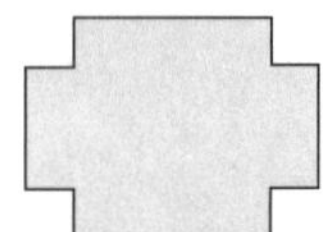

d 

e 

3 Write down the order of rotational symmetry for each of the symbols.

a 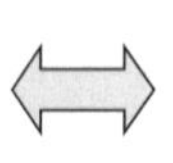

b 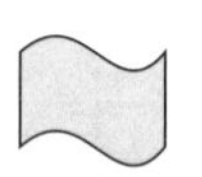

c 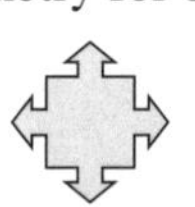

d 

e 

4 The capital letter A fits exactly onto itself only once. So, its order of rotational symmetry is 1. This means that it has no rotational symmetry. Copy these capital letters of the alphabet and write the order of rotational symmetry below each one.

a E **b** H **c** I **d** L **e** N

f Q **g** S **h** Z

★5 Draw two copies of the diagram below.

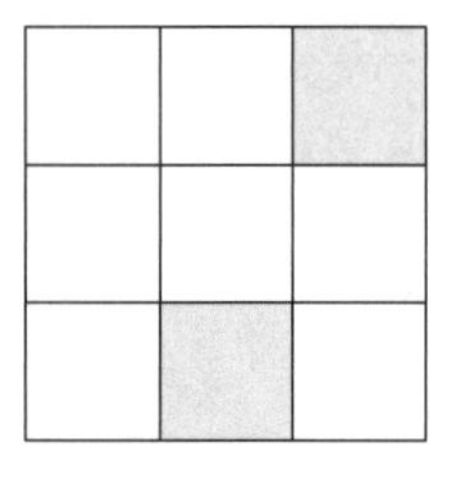

- **a** Shade in two more squares so that the diagram has rotational symmetry of order 2 and no lines of symmetry.
- **b** Shade in two more squares so that the diagram has rotational symmetry of order 1 and exactly 1 line of symmetry.

HOMEWORK 14C

1 Find the number of planes of symmetry in each of these 3-D shapes.

a

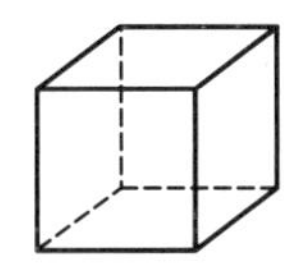

b

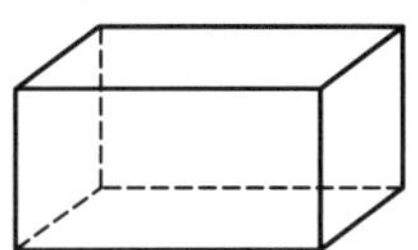

c

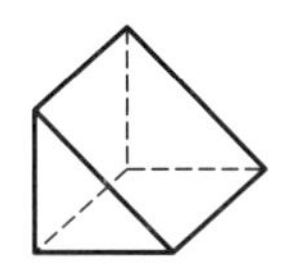

d

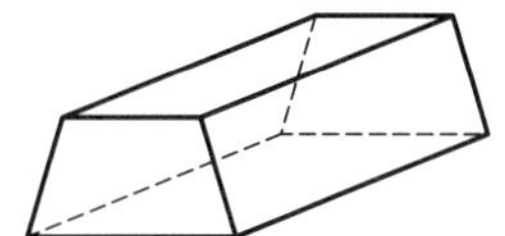

2 This 3-D shape has two planes of symmetry. Draw diagrams to show where they are.

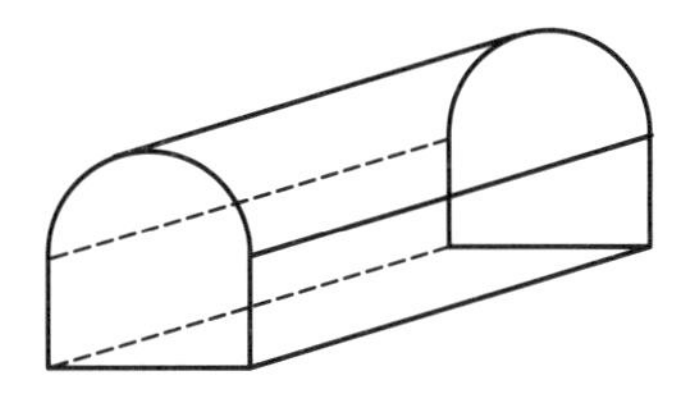

3 The diagram shows half of a 3-D shape. Draw the complete shape so that the shaded part forms a plane of symmetry. What name do we give to this 3-D shape?

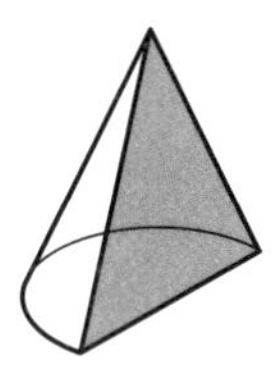

4 How many planes of symmetry does each of the following have?

a

b

c

d

15 Angles

HOMEWORK 15A

1 Use a protractor to find the size of each marked angle.

a

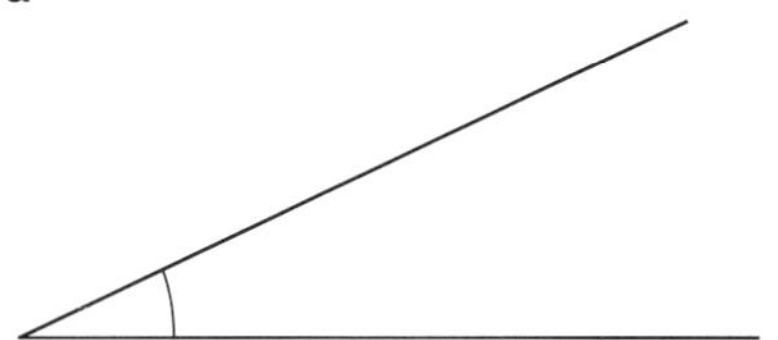

b

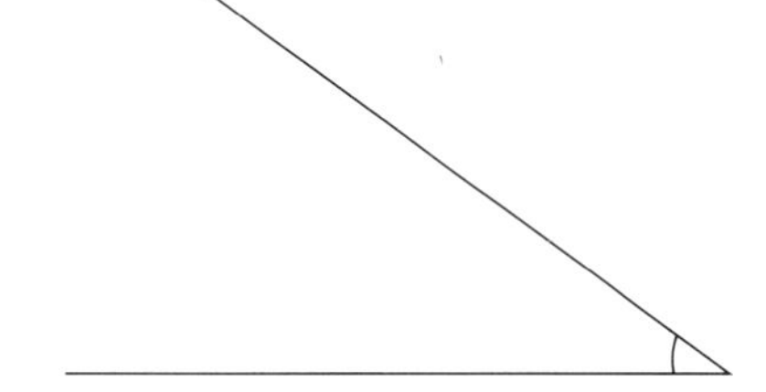

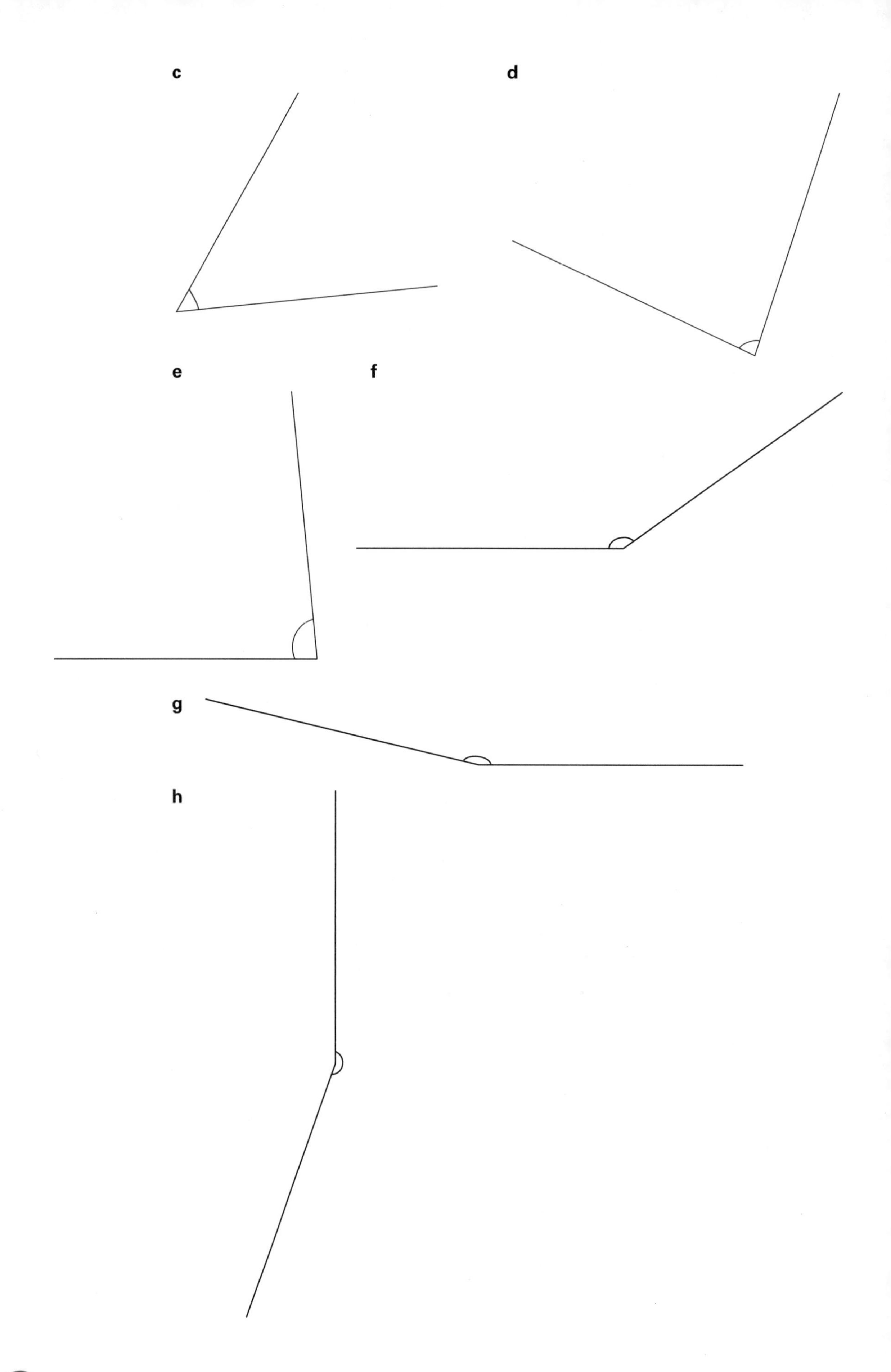
c
d
e
f
g
h

2 Draw angles of the following sizes.

a 30° **b** 42° **c** 55° **d** 68° **e** 75° **f** 140°
g 164° **h** 245°

3 **a** Draw any three acute angles.
b Estimate their sizes. Record your results.
c Measure the angles. Record your results.
d Work out the difference between your estimate and your measurement for each angle.

HOMEWORK 15B

Example Find the value of x in the diagram.

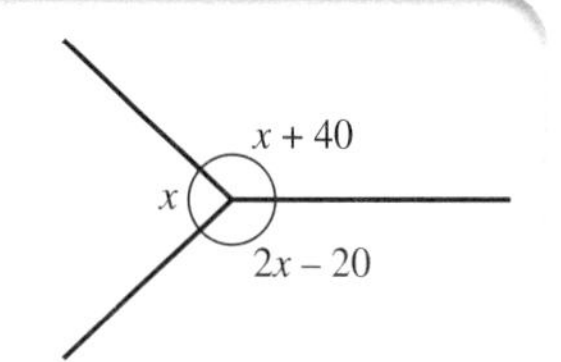

These angles are around a point and add up to 360°.

So $x + x + 40 + 2x - 20 = 360°$

$4x + 20 = 360°$

$4x = 340°$

$x = 85°$

1 Calculate the size of the angle marked with a letter in each of these examples.

a

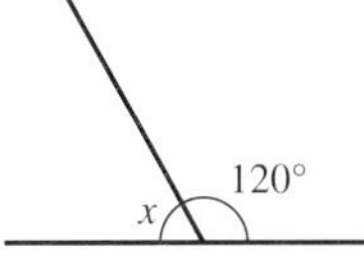

b

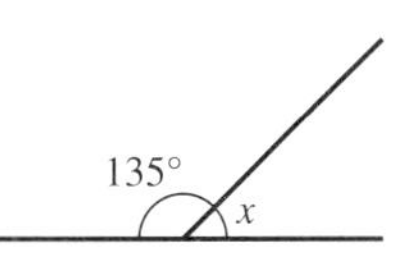

c

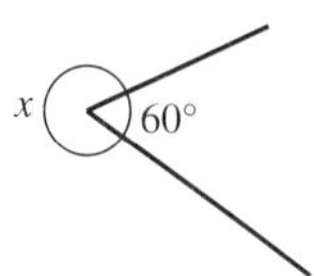

d

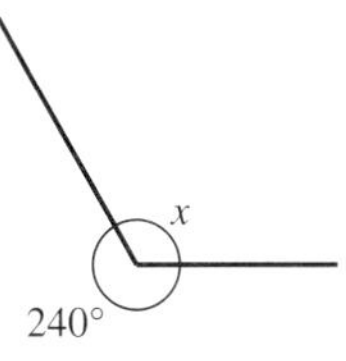

e

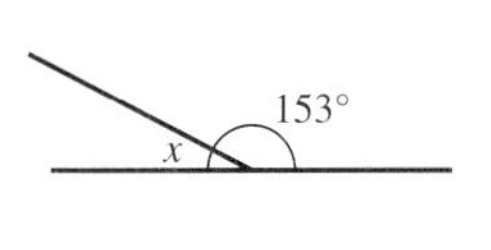

f

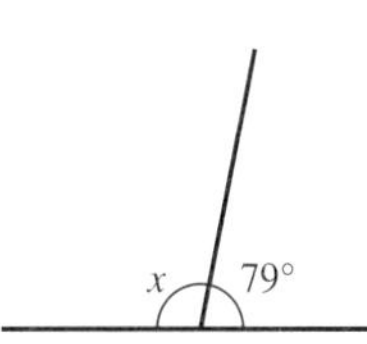

g

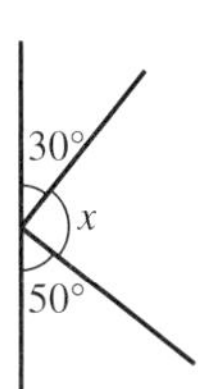

h

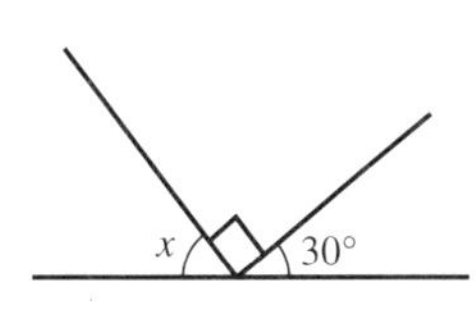

i

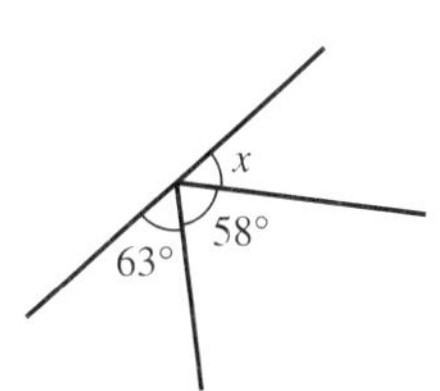

j

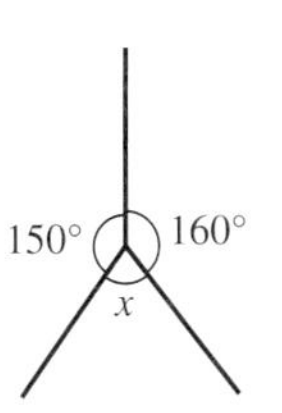

k

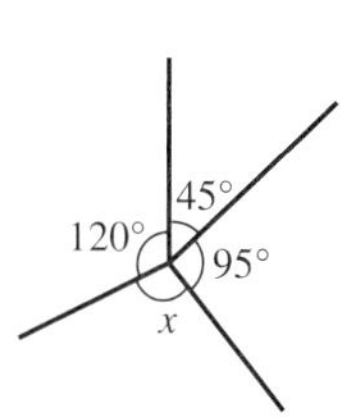

l

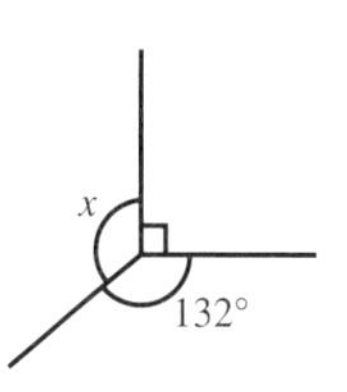

2 Calculate the value of x in each of these examples.

a

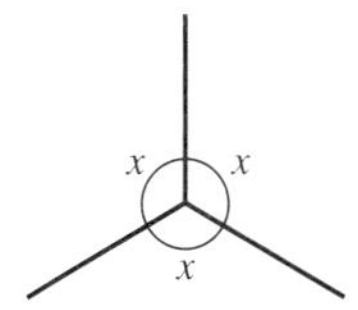

b

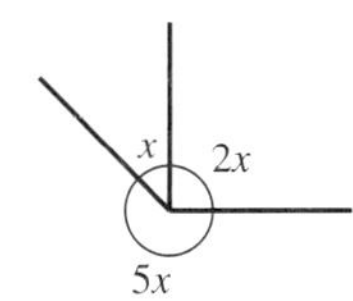

c

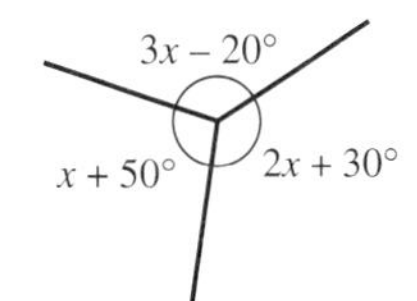

3 Calculate the value of x in each of these examples.

a

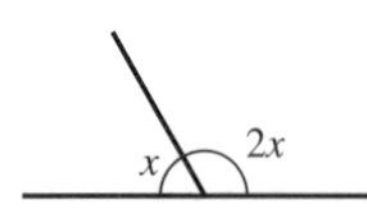

b

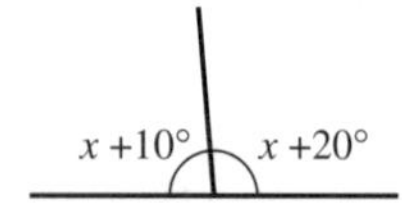

c

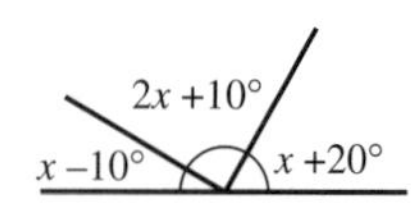

4 Calculate the value of x first and then find the size of angle y in each of these examples.

a

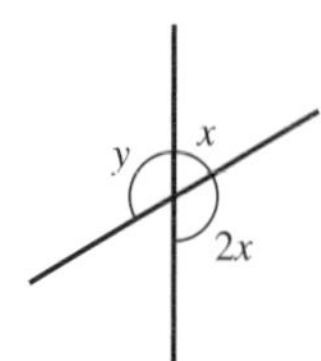

b

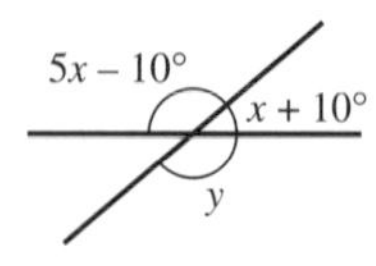

c

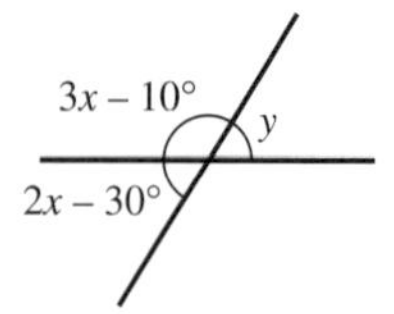

HOMEWORK 15C

1 Find the size of the angle marked with a letter in each of these triangles.

a

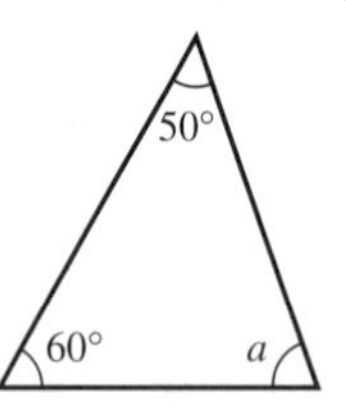

b

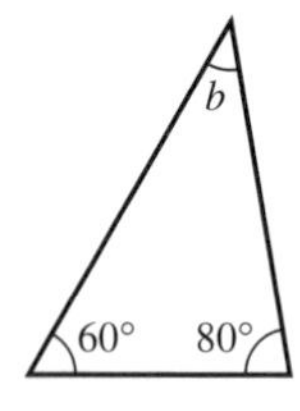

c

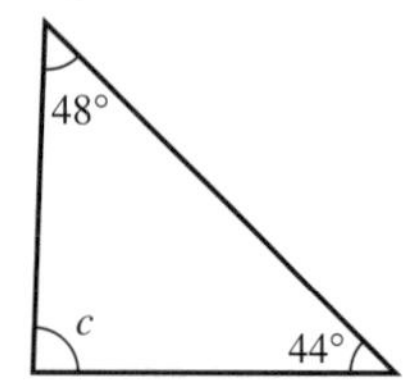

d

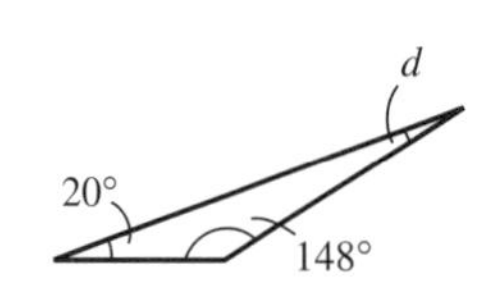

e

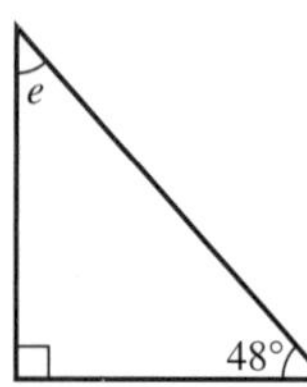

f

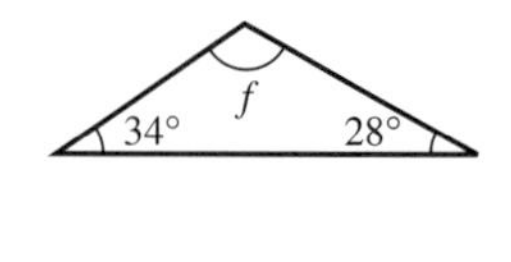

2 State whether each of these sets of angles are the three angles of a triangle? Explain your answers.

a 15°, 85° and 80° **b** 40°, 60° and 90° **c** 25°, 25° and 110°
d 40°, 40° and 100° **e** 32°, 37° and 111° **f** 61°, 59° and 70°

3 The three interior angles of a triangle are given in each case. Find the angle indicated by a letter.

a 40°, 70° and a° **b** 60°, 60° and b° **c** 80°, 90° and c°
d 65°, 72° and d° **e** 130°, 45° and e° **f** 112°, 27° and f°

4 In a triangle all the interior angles are the same.

a What size is each angle?
b What is the special name of this triangle?
c What is special about the sides of this triangle?

5 In the triangle on the right, two of the angles are the same.

a Work out the size of the lettered angles.
b What is the special name of a triangle like this?
c What is special about the sides AB and AC of this triangle?

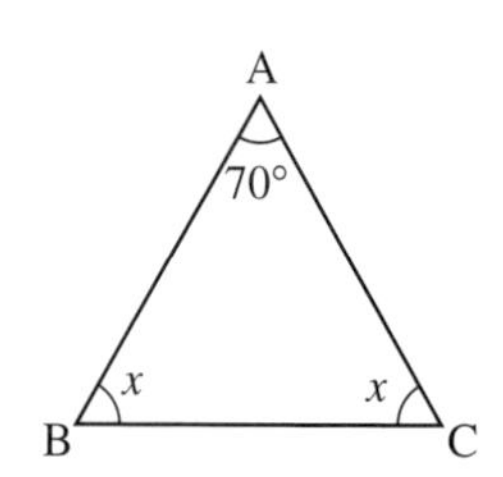

6 In the triangle on the right, the size of the angle at C is twice the size of the angle at A. Work out the size of the lettered angles.

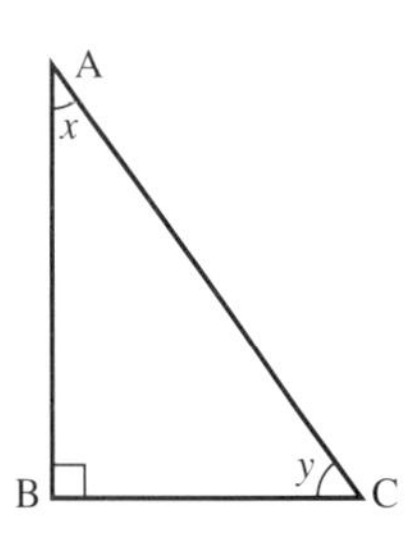

★7 Find the size of the angle marked with a letter in each of the diagrams.

a

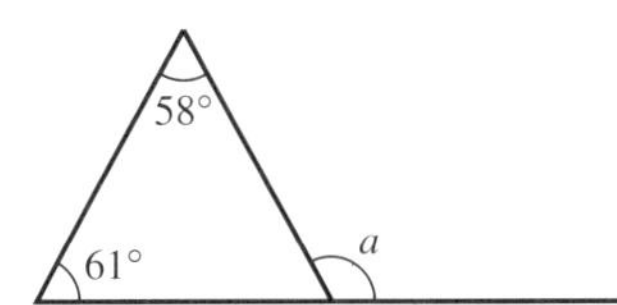

b

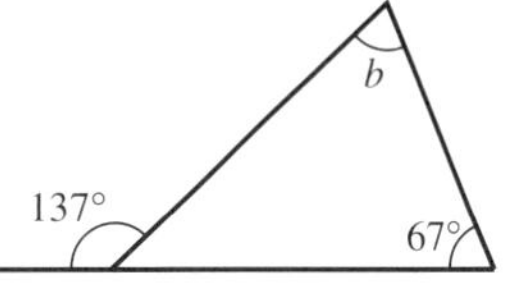

HOMEWORK 15D

1 Find the size of the angle marked with a letter in each of these quadrilaterals.

a

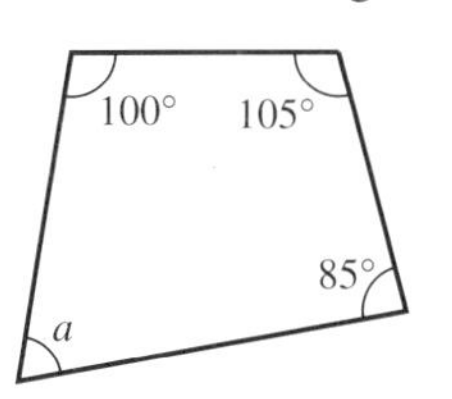

b

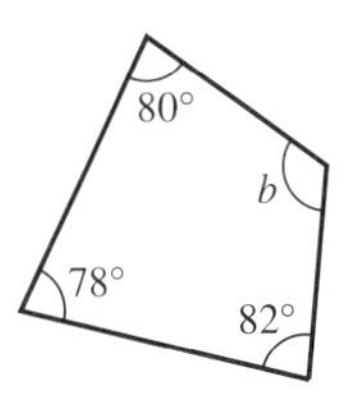

c

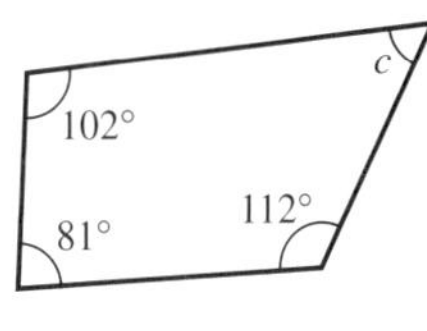

d

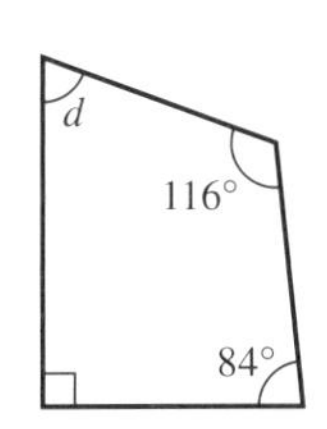

e

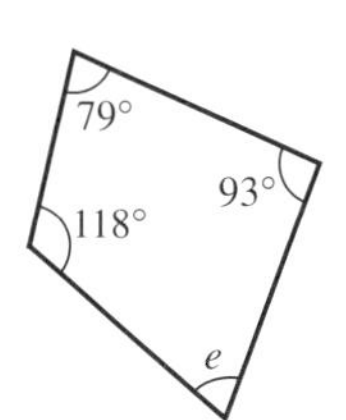

f

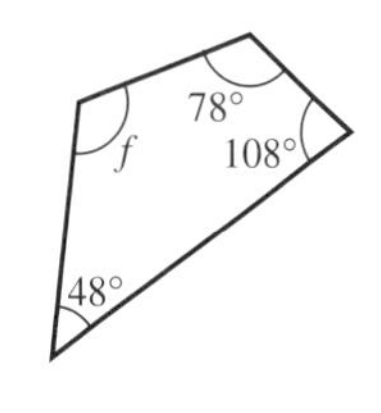

2 State whether each of these sets of angles are the four interior angles of a quadrilateral? Explain your answers.

a 125°, 65°, 70° and 90°

b 100°, 60°, 70° and 130°

c 85°, 95°, 85° and 95°

d 120°, 120°, 70° and 60°

e 112°, 68°, 32° and 138°

f 151°, 102°, 73° and 34°

3 Three interior angles of a quadrilateral are given. Find the fourth angle indicated by a letter.

a 110°, 90°, 70° and a°

b 100°, 100°, 80° and b°

c 60°, 60°, 160° and c°

d 135°, 122°, 57° and d°

e 125°, 142°, 63° and e°

f 102°, 72°, 49° and f°

★4 For the quadrilateral on the right:

a find the size of angle x.

b What type of angle is x?

c What is the special name of a quadrilateral like this?

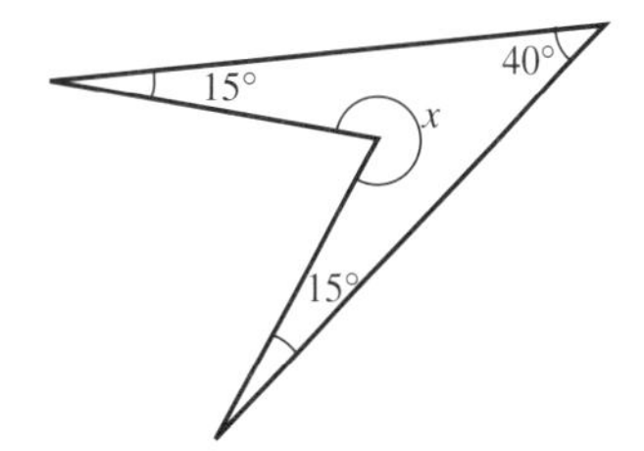

★5 **a** Draw a diagram to explain why the sum of the interior angles of any pentagon is 540°.

b Find the size of the angle x in the pentagon.

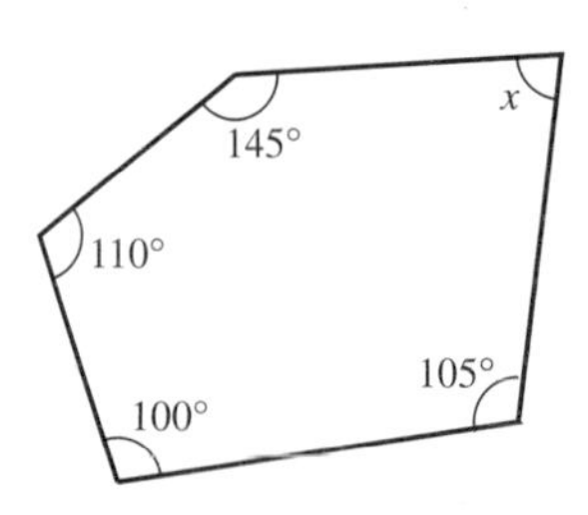

6 Calculate the size of the angle marked with a letter in each of the polygons below.

a

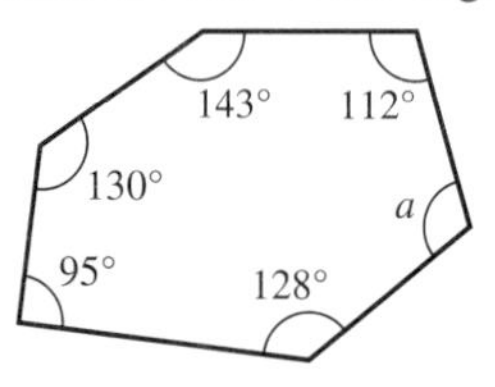

b

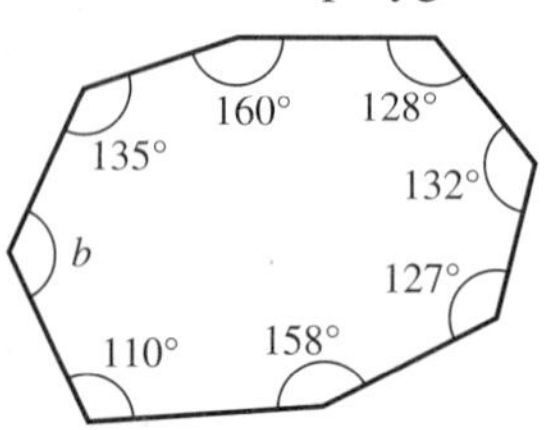

HOMEWORK 15E

1 For each regular polygon below, find the interior angle x and the exterior angle y.

a

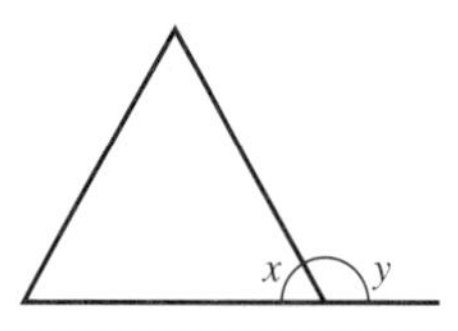

b

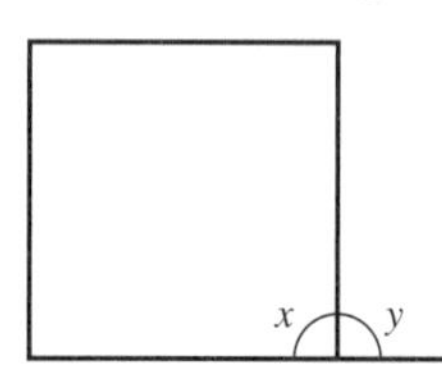

c

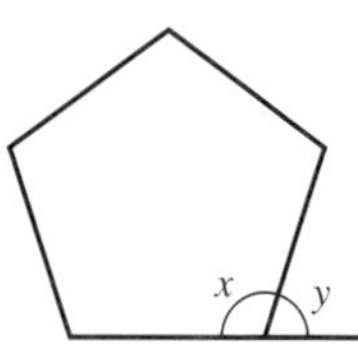

d

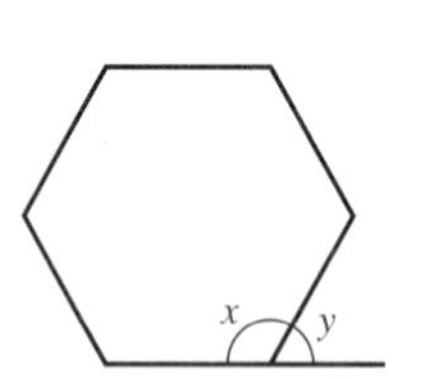

e

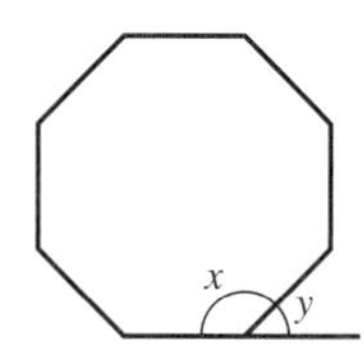

2 Find the number of sides of the regular polygon with an exterior angle of:

a 20° **b** 30° **c** 18° **d** 4°.

3 Find the number of sides of the regular polygon with an interior angle of:

a 135° **b** 165° **c** 170° **d** 156°.

4 What is the name of the regular polygon whose interior angle is treble its exterior angle?

HOMEWORK 15F

1 State the size of the lettered angles in each diagram.

a

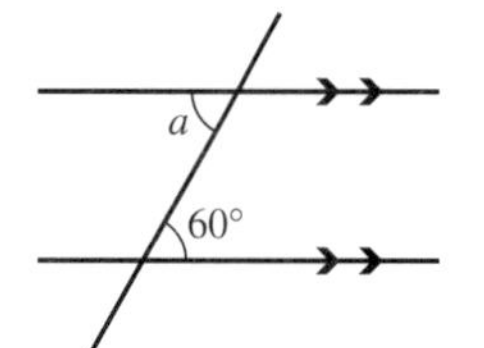

b

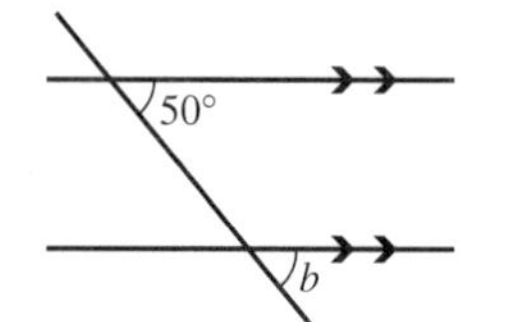

c

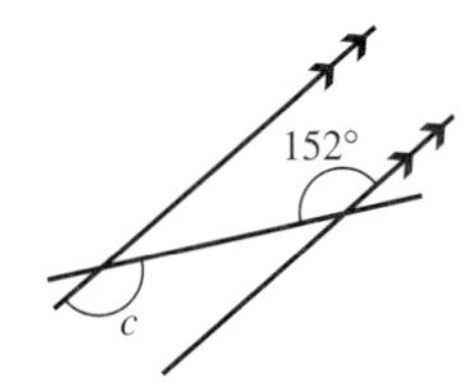

d

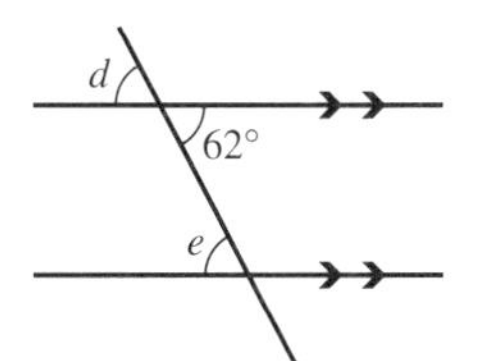

e

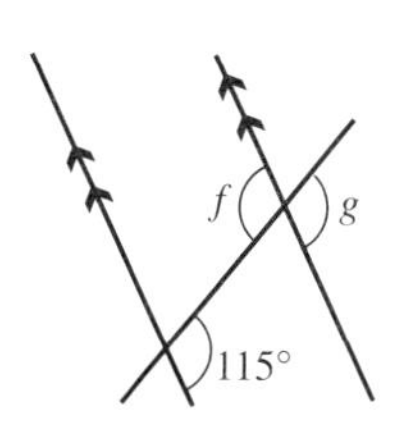

f

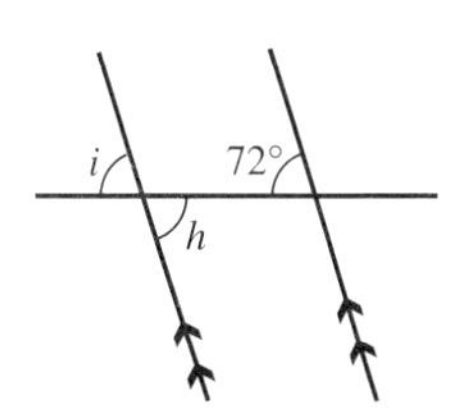

2 State the size of the lettered angles in each diagram.

a

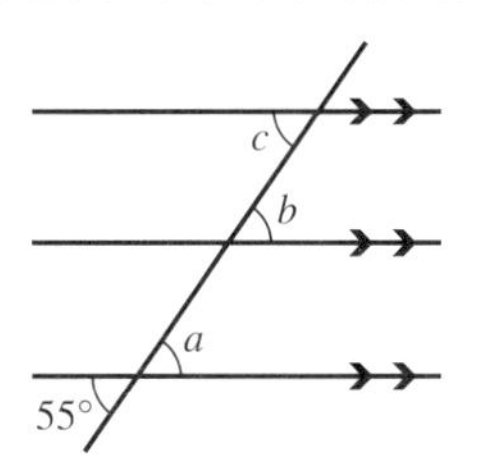

b

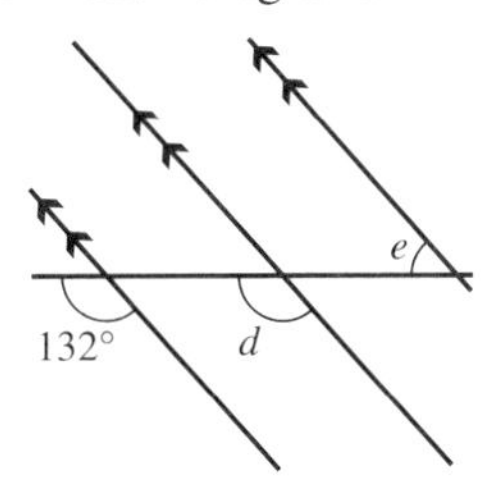

c

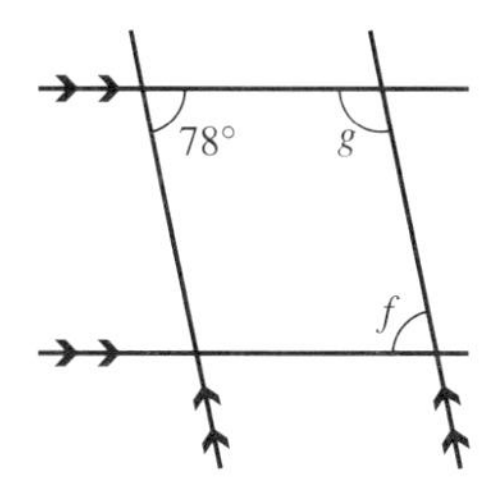

3 State the size of the lettered angles in these diagrams.

a

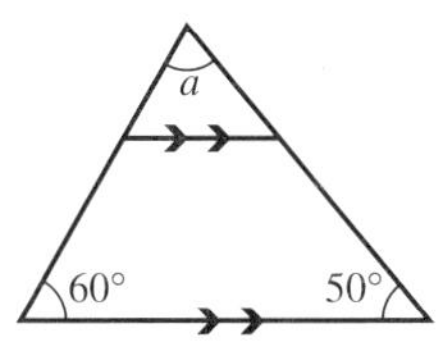

b

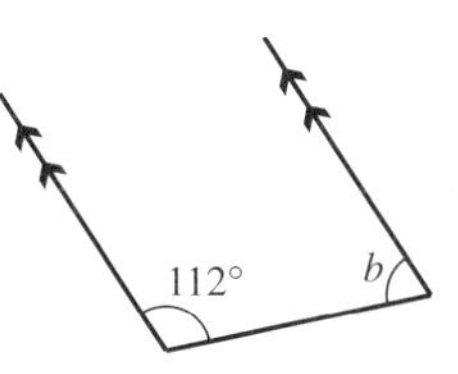

4 Find the values of x and y in these diagrams.

a

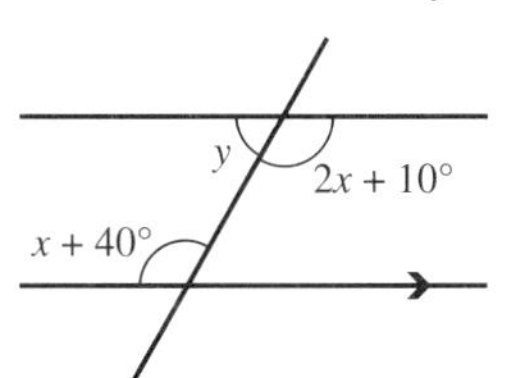

b

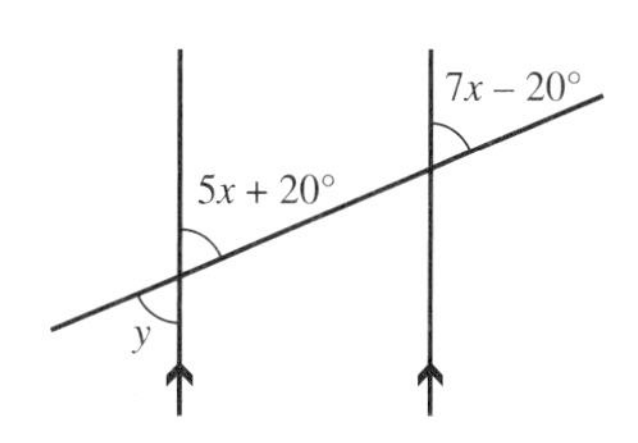

HOMEWORK 15G

1 For each of these trapeziums, calculate the sizes of the lettered angles.

a

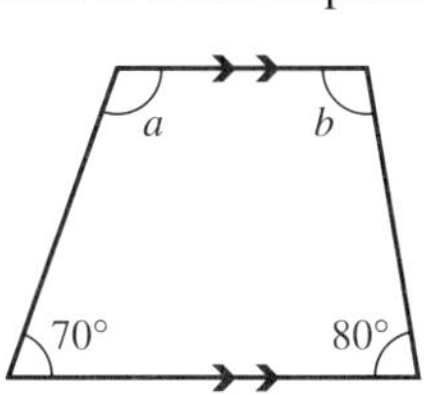

b

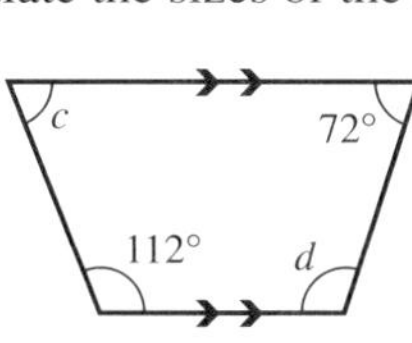

c

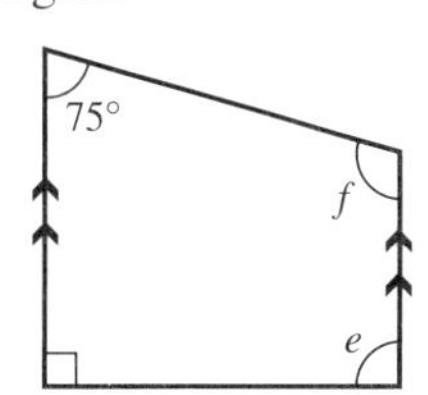

2 For each of these parallelograms, calculate the sizes of the lettered angles.

a

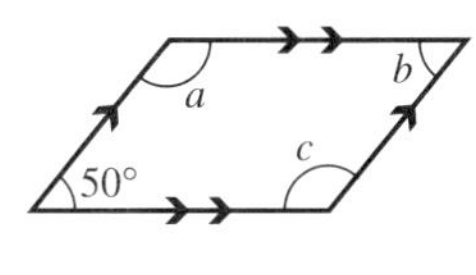

b

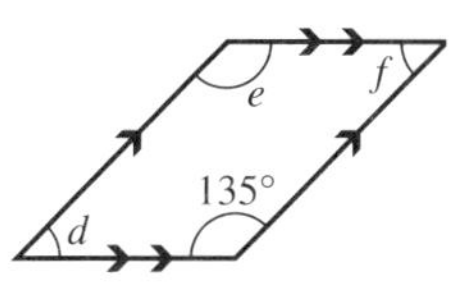

c

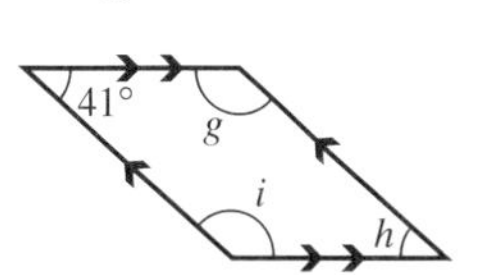

3 For each of these kites, calculate the sizes of the lettered angles.

a

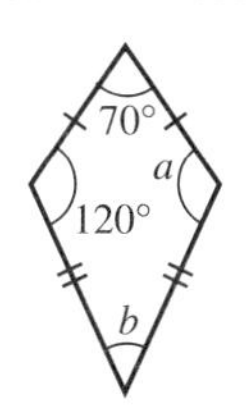

b

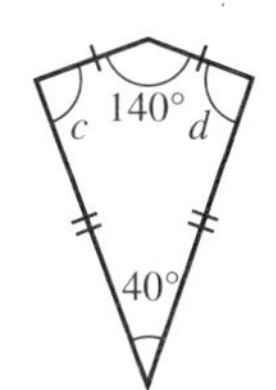

c

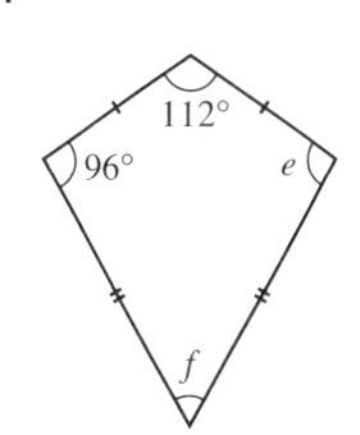

4 For each of these rhombuses, calculate the sizes of the lettered angles.

a

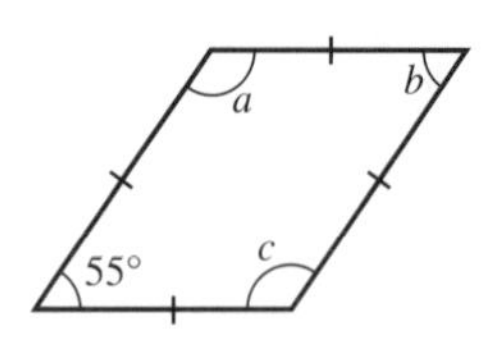

b

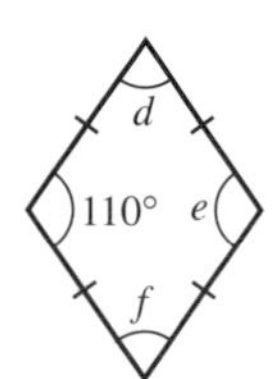

c

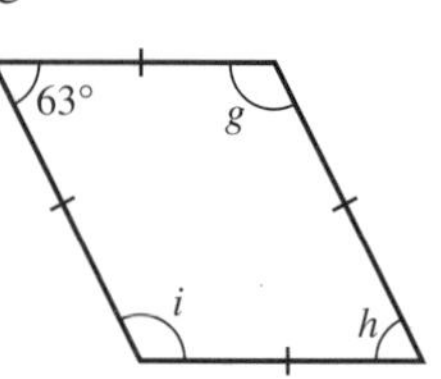

HOMEWORK 15H

1 **a** Write down the bearing of B from A.

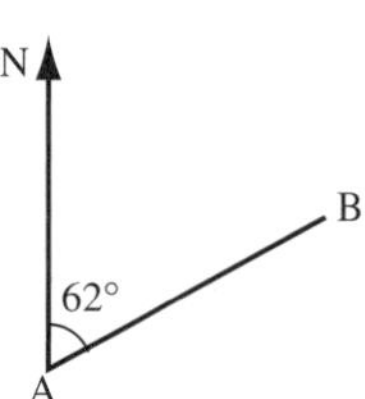

b Write down the bearing of D from C.

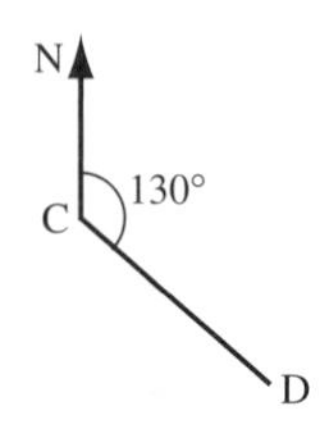

c Write down the bearing of F from E.

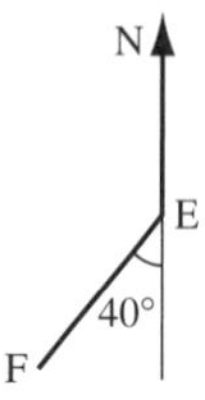

d Write down the bearing of H from G.

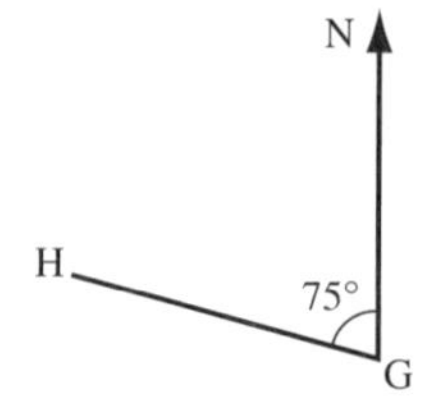

★**2** Look at the map of Britain.
By measuring angles, find the bearing of

a London from Edinburgh
b London from Cardiff
c Edinburgh from Cardiff
d Cardiff from London

16 Circles

HOMEWORK 16A

1 Measure the radius of each of the following circles, giving your answers in centimetres. Write down the diameter of each circle.

a

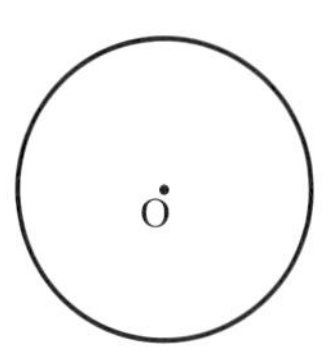

b

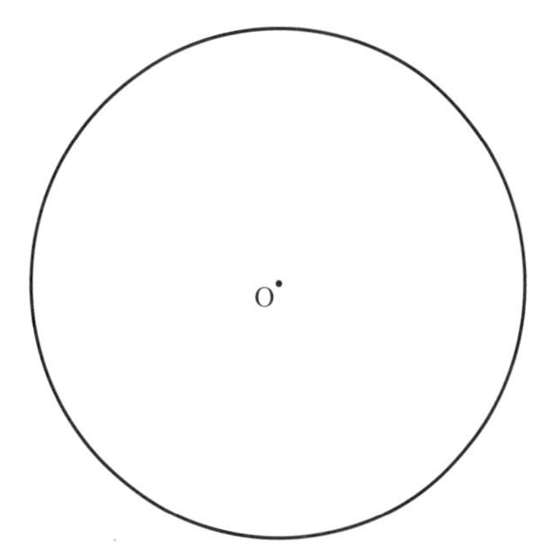

c

2 Draw circles with the following measurements.

a Radius = 1.5 cm **b** Radius = 4 cm

c Diameter = 7 cm **d** Diameter = 9.6 cm

3 Accurately draw the following shapes.

a

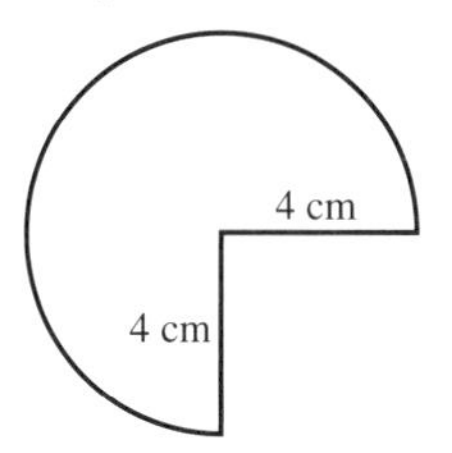

b

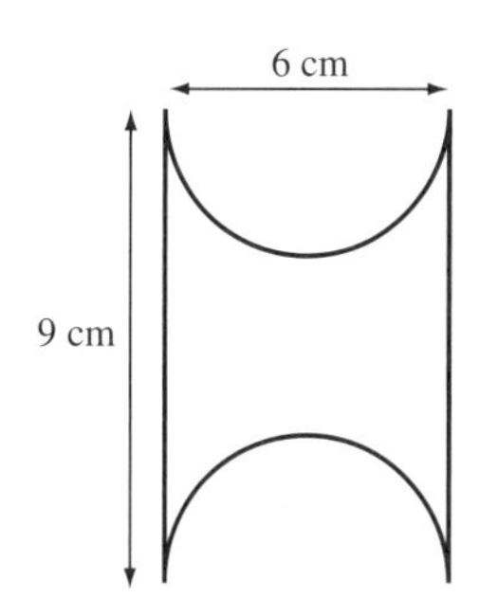

4 Draw an accurate copy of this diagram. What is the length of the diameter of the circle?

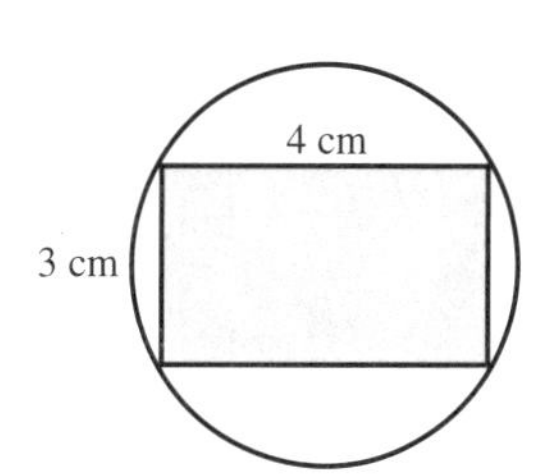

5 Draw an accurate copy of this diagram at full scale. What is the diameter of the circle?

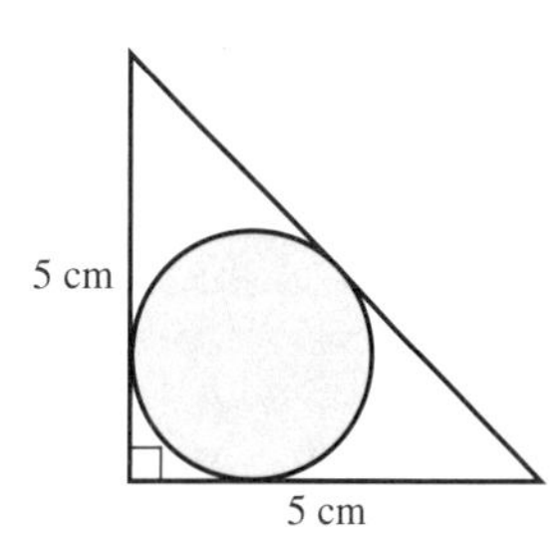

HOMEWORK 16B

Example Calculate the circumference of the circle with a diameter of 4 cm.

Use the formula $c = \pi d$. So $c = \pi \times 4 = 12.6$ cm (1dp).

1 Calculate the circumference of each circle illustrated below.
Give your answers to one decimal place.

a

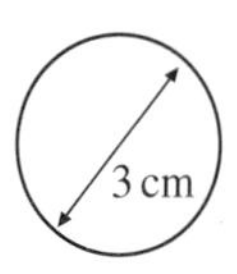

b

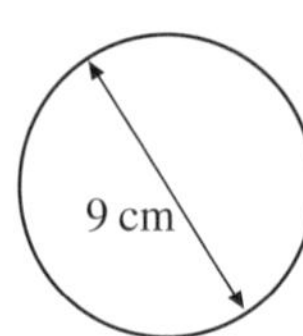

c

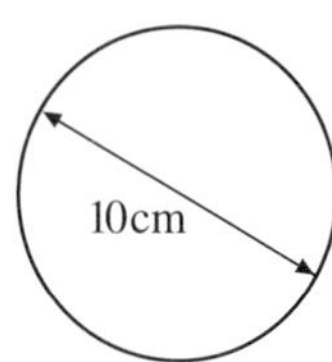

d

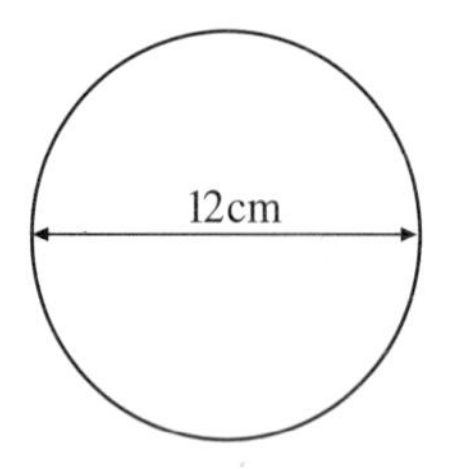

e

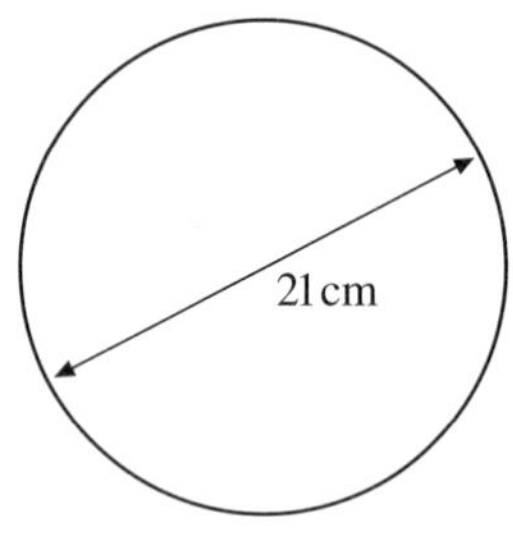

2 Calculate the circumference of each circle illustrated below.
Give your answers to one decimal place.

a

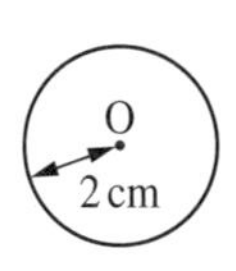

b

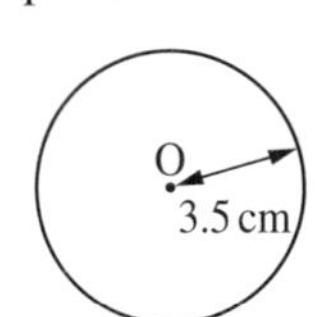

c

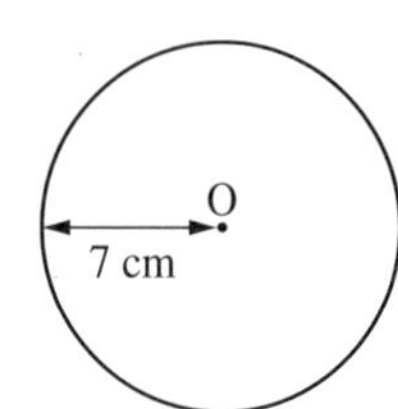

d

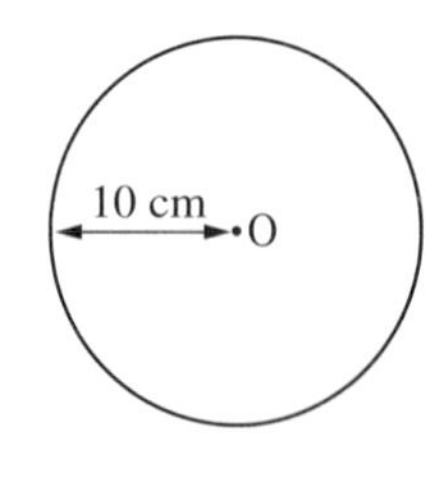

e

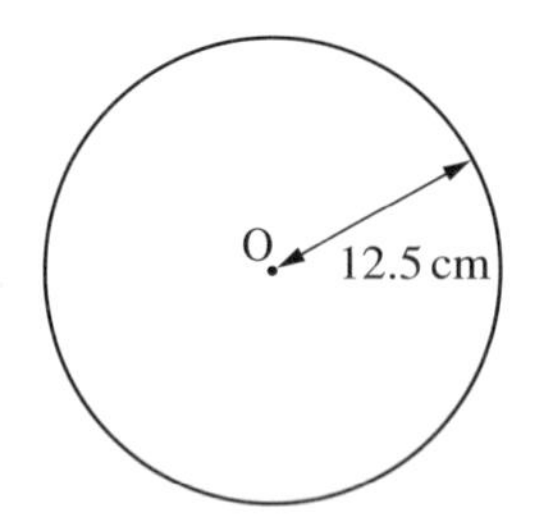

★**3** A fence is to be put around a circular pond which has a diameter of 15 m. What is the length of fencing required, if the fencing is bought in 1 m lengths?

★**4** Roger trains for an athletics competition by running round a circular track which has a radius of 50 m.

a Calculate the circumference of the track. Give your answer to 1 decimal place.

b How many complete circuits will he need to run to be sure of running 5000 m?

★**5** Calculate the perimeter of this semicircle.

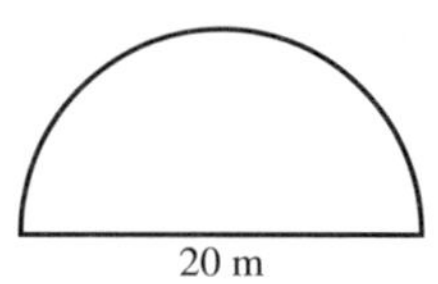

★**6** What is the diameter of a circle whose circumference is 40 cm? Give your answer to one decimal place.

Example Calculate the area of a circle with a radius of 7 cm.

Use the formula $A = \pi r^2$. So $A = \pi \times r \times r = \pi \times 7 \times 7 = 153.9\ \text{cm}^2$ (1dp).

1 Calculate the area of each circle illustrated below. Give your answers to one decimal place.

a

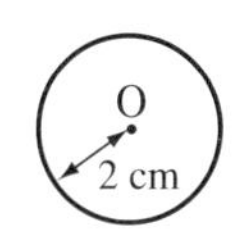

b

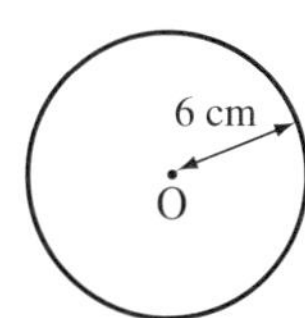

c

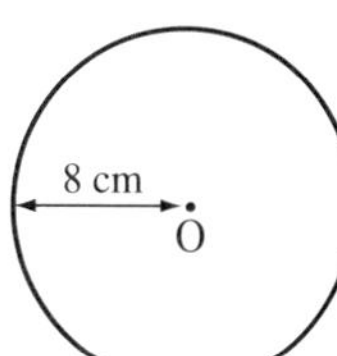

d

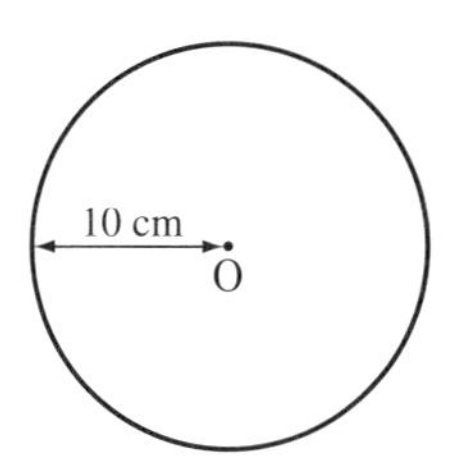

e

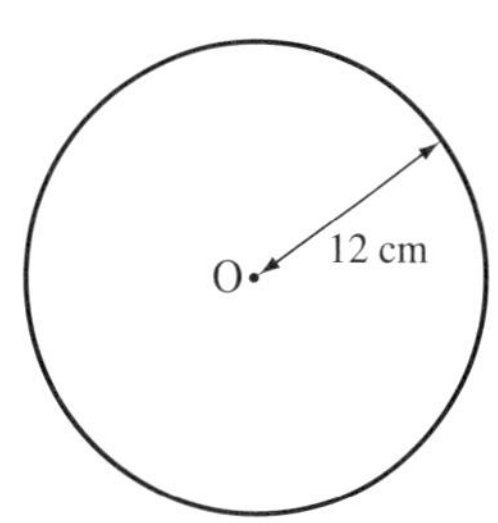

2 Calculate the area of each circle illustrated below. Give your answers to one decimal place.

a

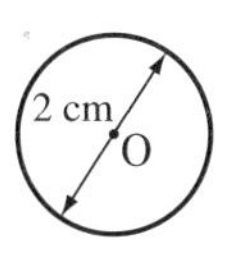

b

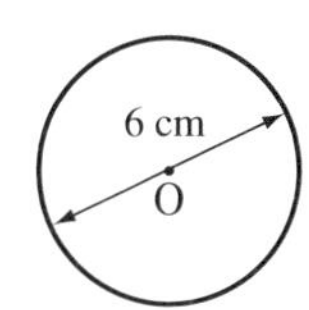

c

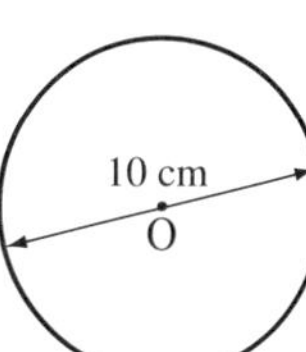

d

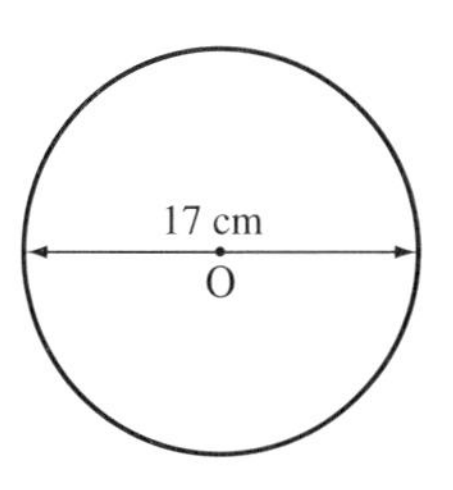

e

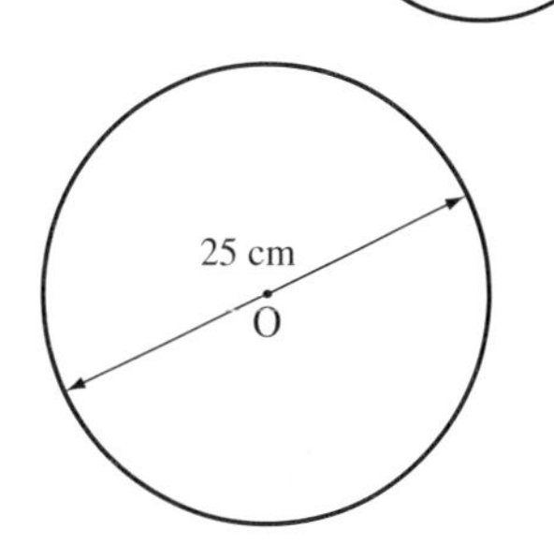

★**3** A circular table has a diameter of 80 cm.

a Calculate the circumference of the table, giving your answer in metres to one decimal place.

b Calculate the area of the table, giving your answer in square metres to one decimal place.

★**4** The diagram shows a circular path around a flower bed in a park. The radius of the flower bed is 6 m and the width of the path is 1 m.

a Calculate the area of the flower bed.

b Write down the radius of the large circle.

c Calculate the area of the large circle.

d Calculate the area of the path.

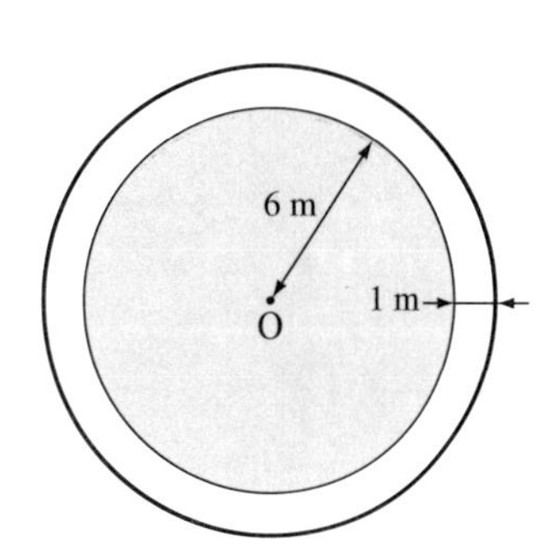

★5 The diagram shows a running track.

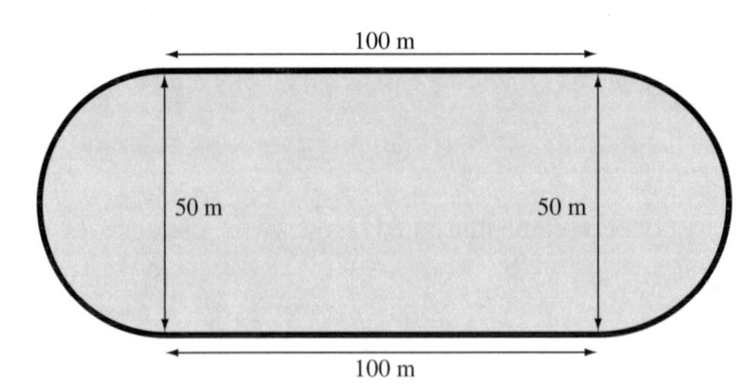

a Calculate the perimeter of the track. Give your answer to the nearest whole number.

b Calculate the total area inside the track. Give your answer to the nearest whole number.

6 A circle has a circumference of 50 cm.

a Calculate the diameter of the circle to one decimal place.

b What is the radius of the circle to one decimal place?

c Calculate the area of the circle to one decimal place.

HOMEWORK 16D

Leave all your answers in terms of π.

1 State the circumference of the following circles.

a Diameter 7 cm **b** Radius 5 cm **c** Diameter 19 cm **d** Radius 3 cm

2 State the area of the following circles.

a Radius 8 cm **b** Diameter 7 cm **c** Diameter 18 cm **d** Radius 9 cm

3 State the diameter of a circle with a circumference of 4π cm.

4 State the radius of a circle with an area of $25\pi\ cm^2$.

5 State the diameter of a circle with a circumference of 20 cm.

6 State the radius of a circle with an area of $20\ cm^2$.

★7 Calculate **i** the perimeter and **ii** the area for each of the following shapes, giving your answers in terms of π.

a

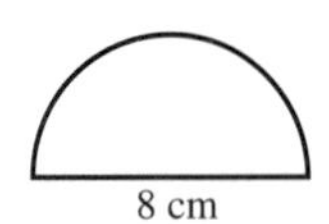

b

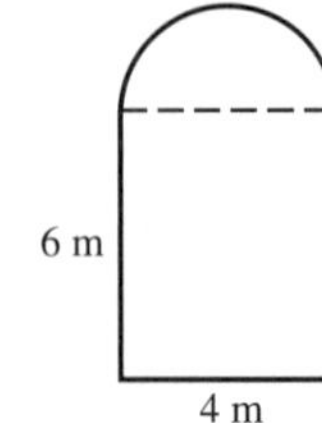

17 Scale and drawing

HOMEWORK 17A 

1 Read the values from the following scales. Remember to state the units.

a

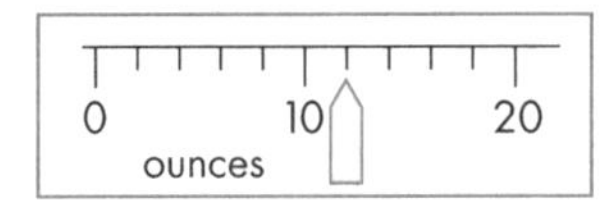

b

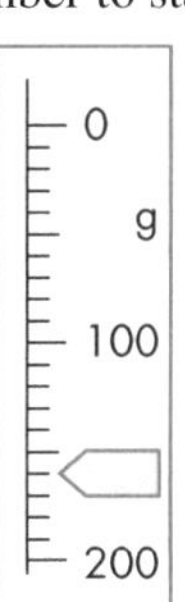

c

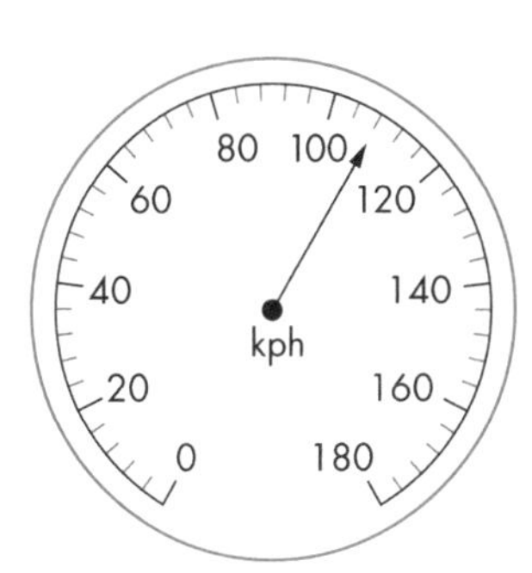

d

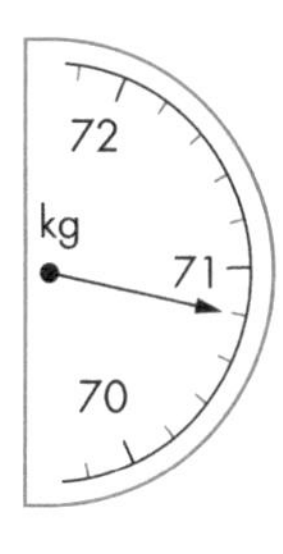

2 Read the temperatures shown on each of these thermometers.

a

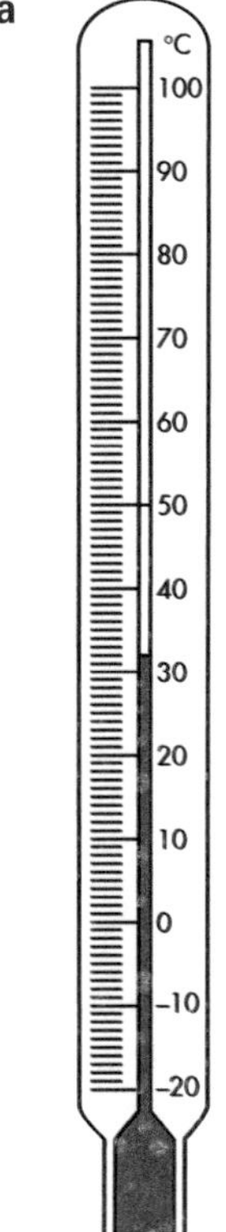

b

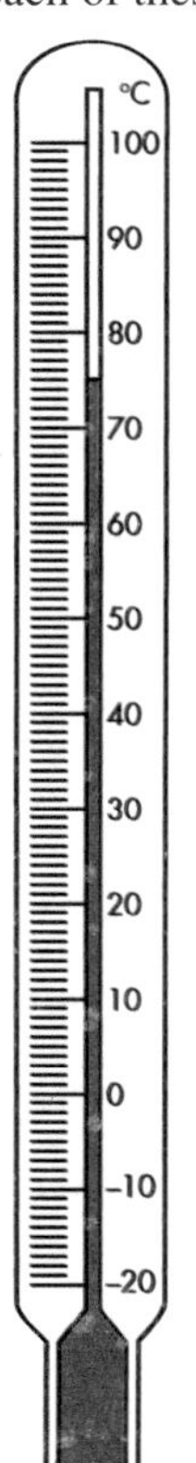

c

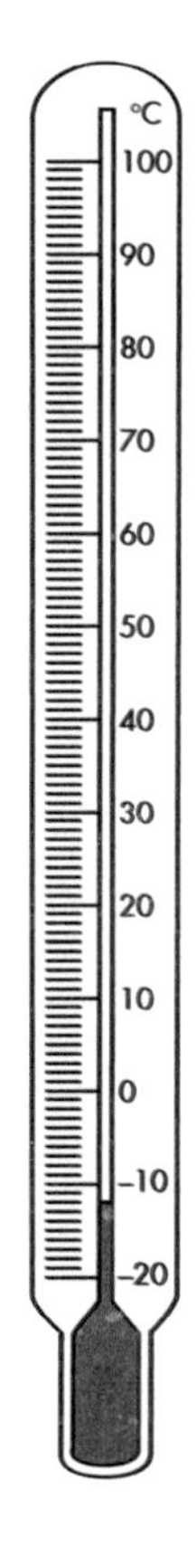

3 Read the values shown on these scales.

a

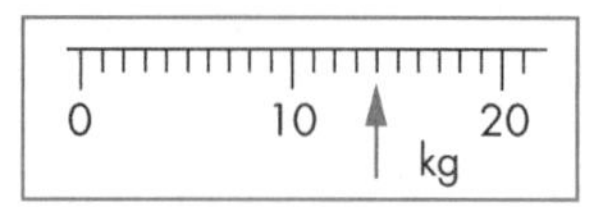

b

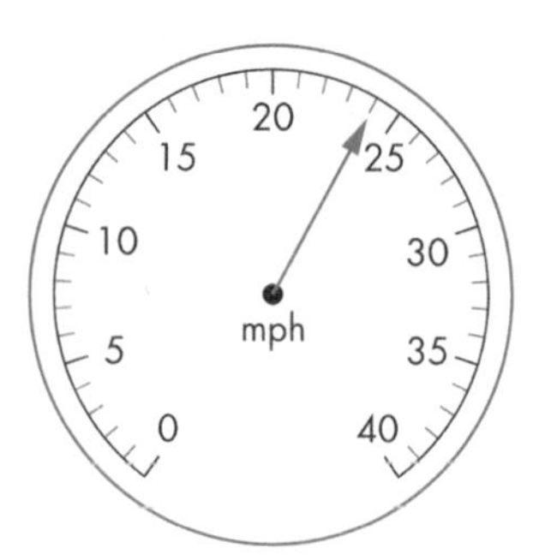

c

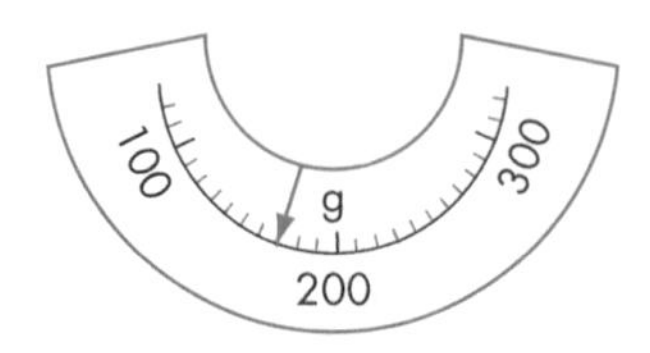

d

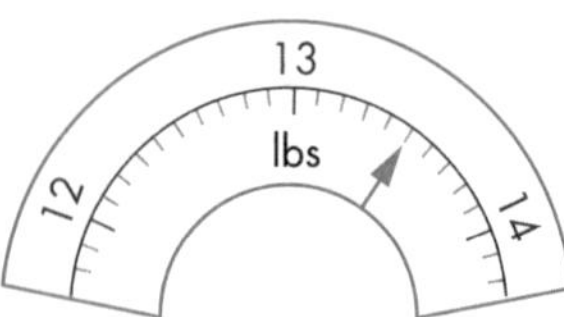

4 A boy is standing on some bathroom scales.

a What is the weight of the boy?

The boy now stands on the scales carrying his rucksack.

b How much does the boy's rucksack weigh?

HOMEWORK 17B

1 The picture shows a teacher standing by a whiteboard in a classroom.

a Estimate the height of the door.

b Estimate the length of the classroom.

c Estimate the length and width of the whiteboard.

2 Estimate the height of the giraffe.

3 **a** Estimate the weight of one apple.

b Estimate the weight of one orange.

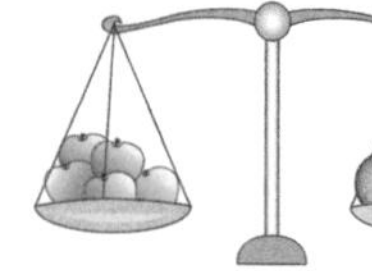

4 The height of The London Eye is 135 m.
Use this information to estimate the height of:

a The Eiffel Tower

b Sears Tower

c Warszawa Radio Mast.

London Eye

Eiffel Tower

Sears Tower

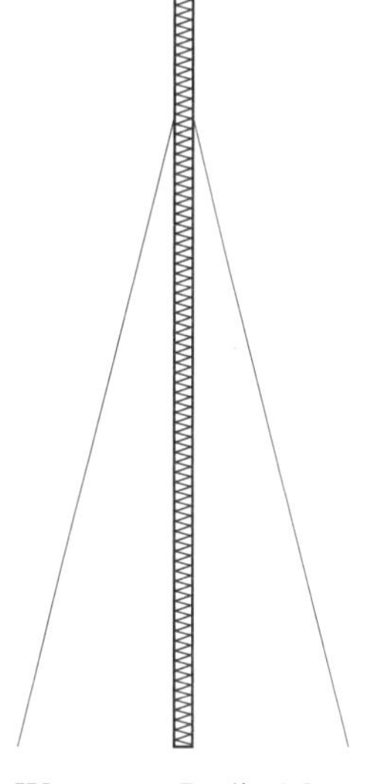
Warszawa Radio Mast

HOMEWORK 17C

1 The grid below shows the floor plan of a kitchen. The scale is 1 cm to 30 cm.

Work space
Cooker
Work space
Sink Unit
Fridge
Door
Cupboards
Door

a State the actual dimensions of:

i the sink unit **ii** the cooker **iii** the fridge **iv** the cupboards.

b Calculate the actual total area of the work space.

★2 Below is a sketch of a ladder leaning against a wall.

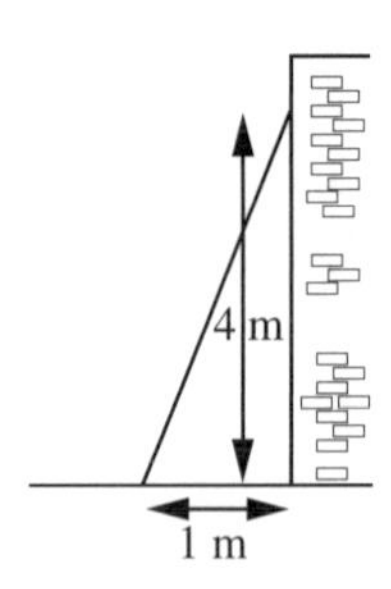

The bottom of the ladder is 1 m away from the wall and it reaches 4 m up the wall.

a Make a scale drawing to show the position of the ladder. Use a scale of 4 cm to 1 m.

b Use your scale drawing to work out the actual length of the ladder.

3 The map below is drawn to a scale of 1 cm to 2 km.

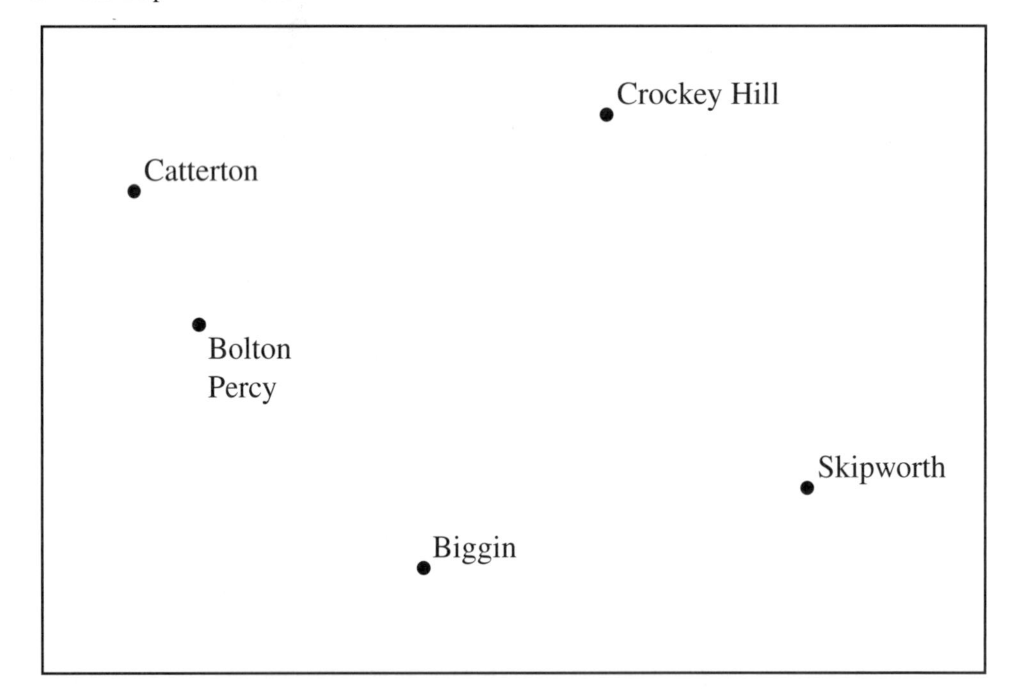

Find the distances between:

a Biggin and Skipworth

b Bolton Percy and Crockey Hill

c Skipworth and Catterton

d Crockey Hill and Biggin

e Catterton and Bolton Percy.

4 The map below shows the position of four fells in the Lake District. The map is drawn to a scale of 1 : 150 000.

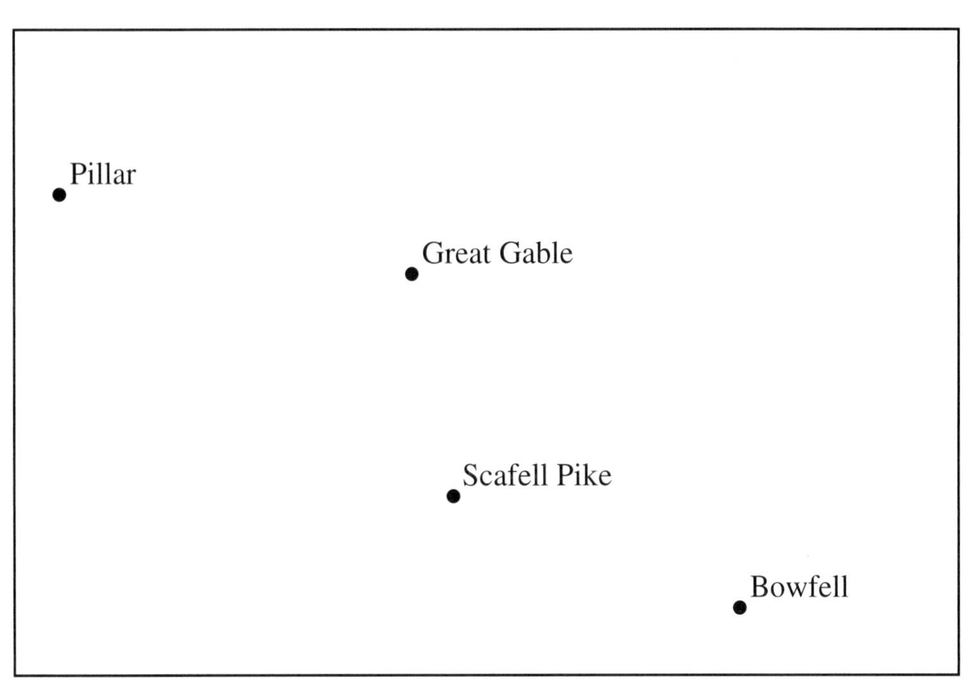

To the nearest half kilometre, find the actual direct distances from:

a Scafell Pike to Great Gable
b Scafell Pike to Pillar
c Great Gable to Pillar
d Pillar to Bowfell
e Bowfell to Great Gable.

HOMEWORK 17D

1 Four nets are shown below. Copy the nets which would make a cube.

a

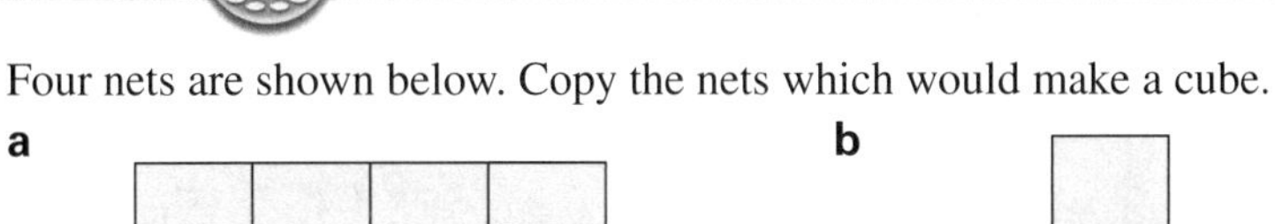

b

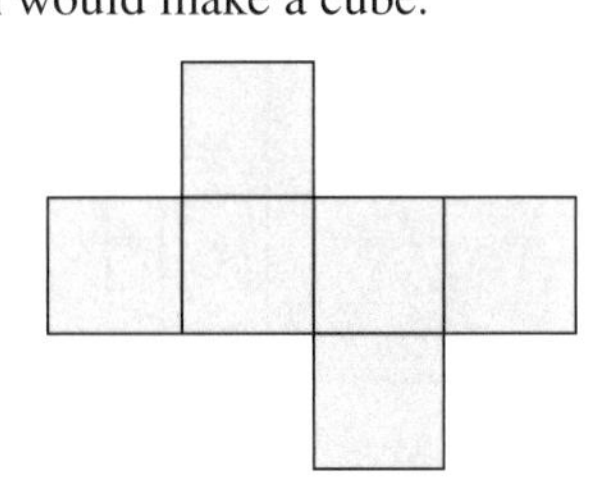

c

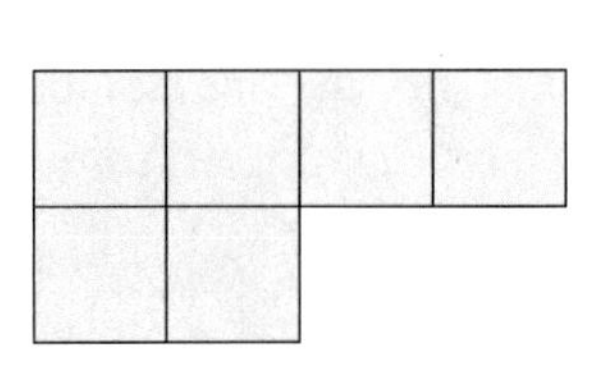

d

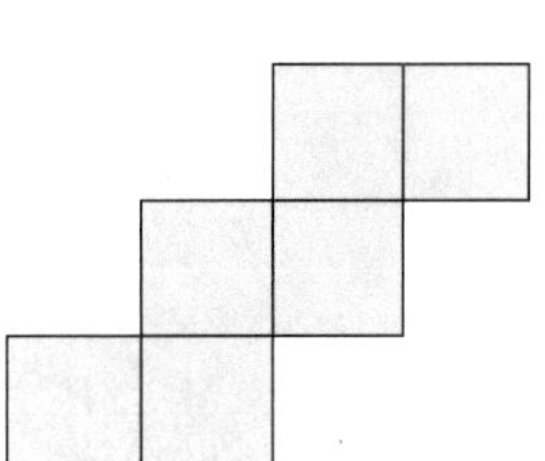

2 Draw, on squared paper, an accurate net for each of these cuboids.

a

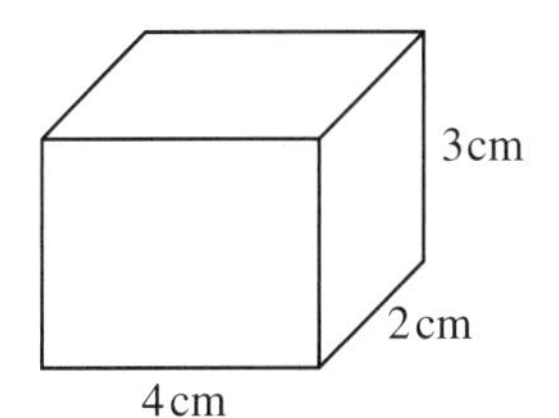

b

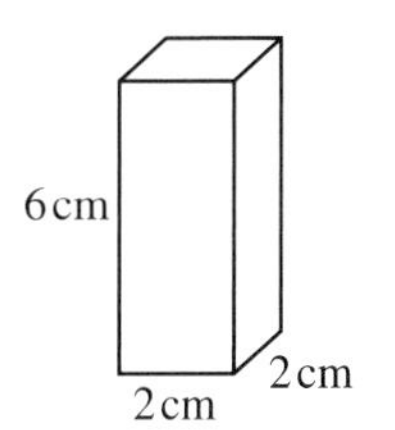

3 Draw an accurate net for this triangular prism.

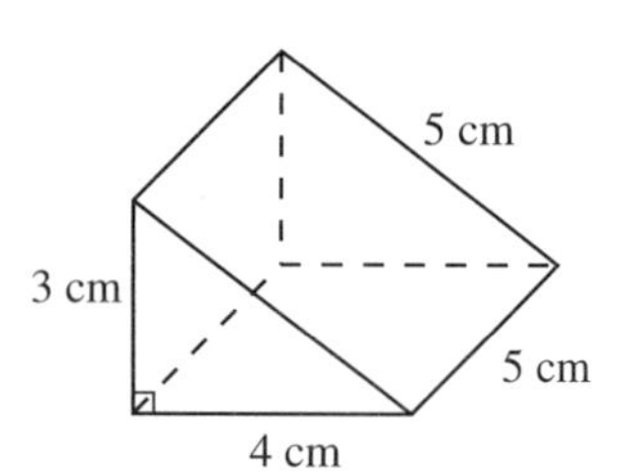

★4 The diagram shows a sketch of a square-based pyramid.

a Write down how many:

i vertices

ii edges

iii faces the pyramid has.

b Draw an accurate net for the pyramid.

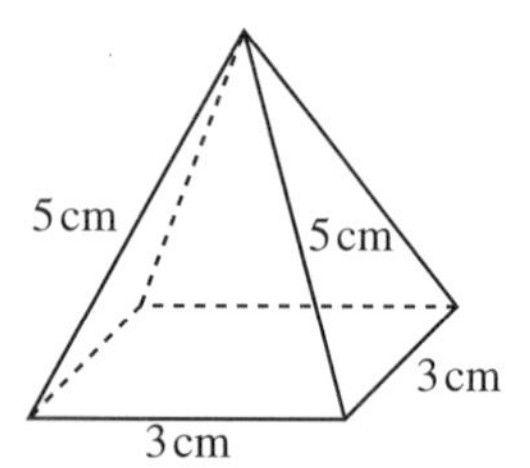

HOMEWORK 17E

1 Draw accurately each of these cuboids on an isometric grid.

a

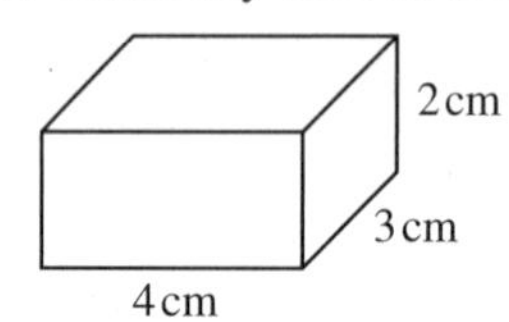

b

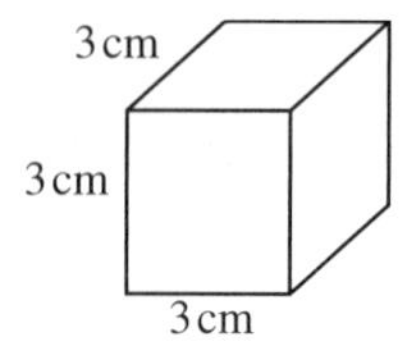

2 Draw accurately each of these 3-D shapes on an isometric grid.

a

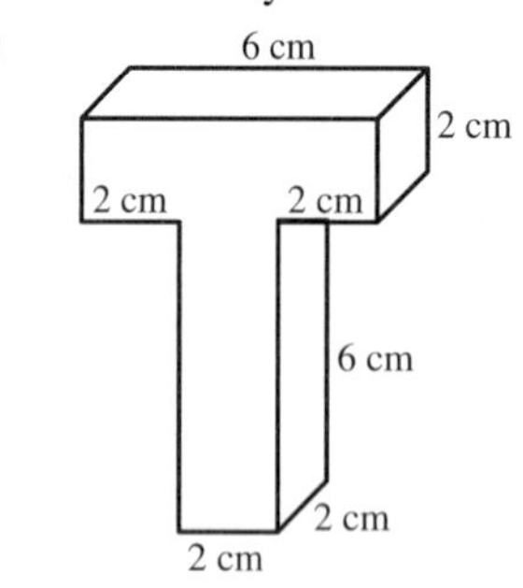

b

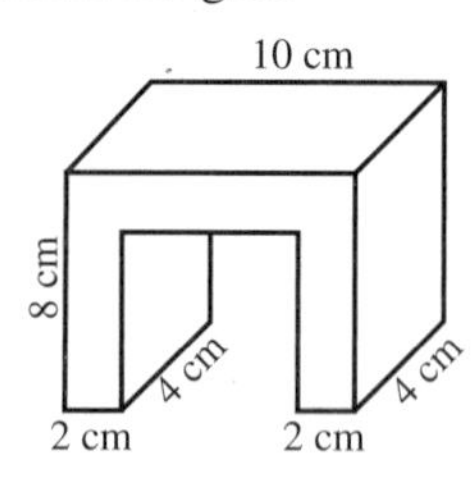

3 For each of the following 3-D shapes, draw on squared paper:

i the plan **ii** the front elevation **iii** the side elevation.

a

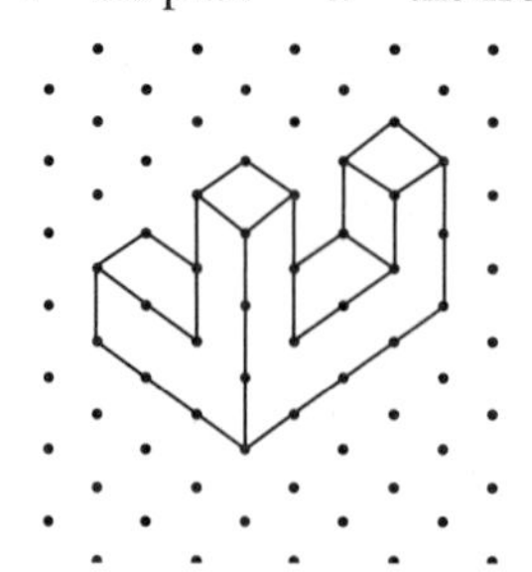

b

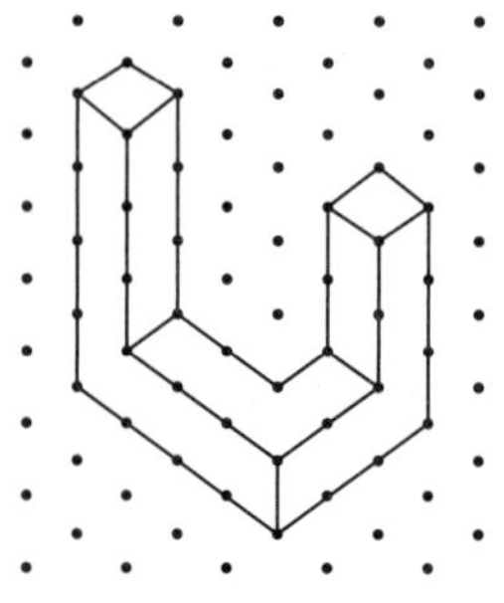

18 Transformations

HOMEWORK 18A

1 State whether each pair of shapes **a** to **f** are congruent or not.

a

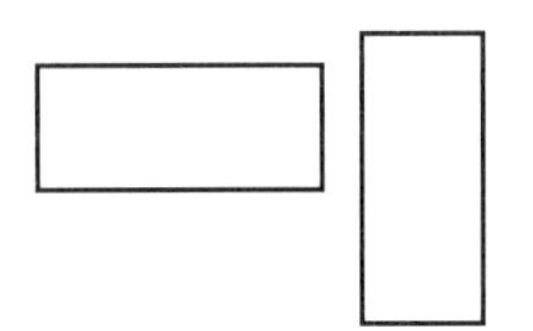

b

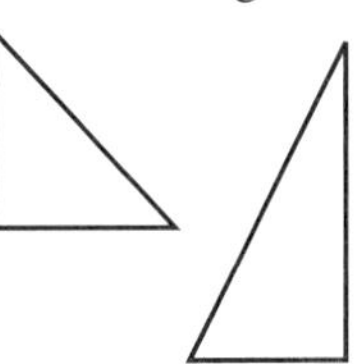

c

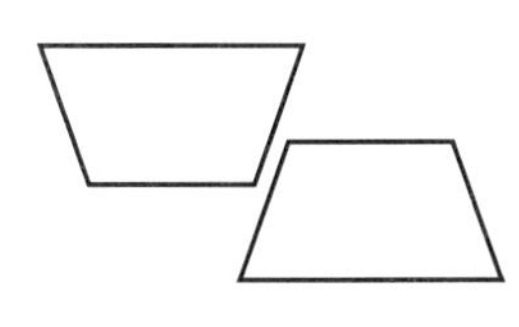

d

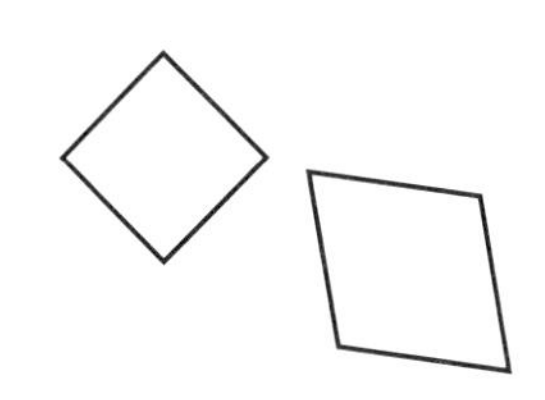

e

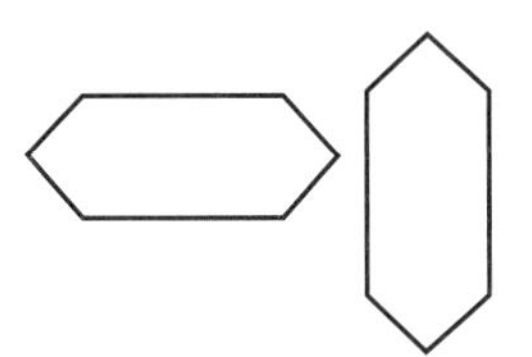

f

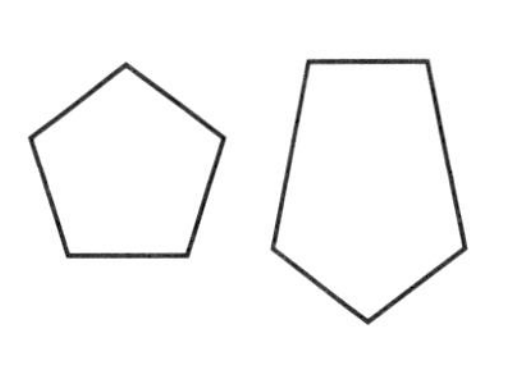

2 Which figure in each group of shapes is not congruent to the other two?

a

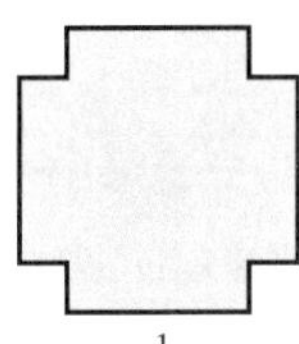

1

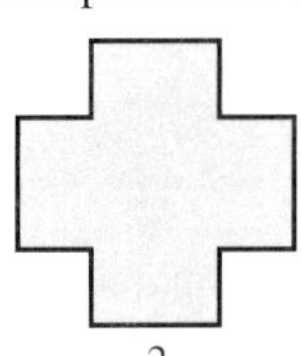

2

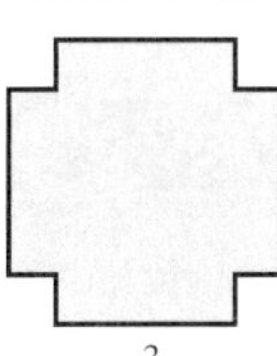

3

b

1

2

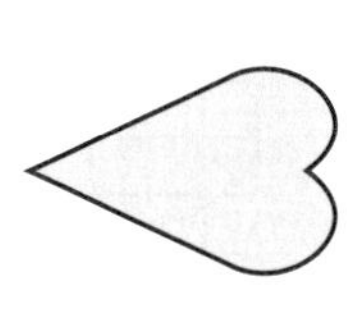

3

c

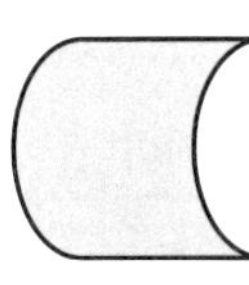

1

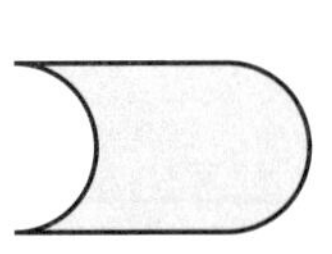

2

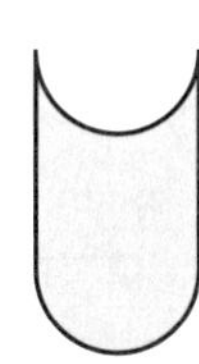

3

d

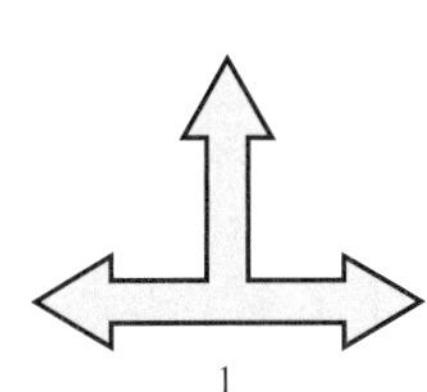

1

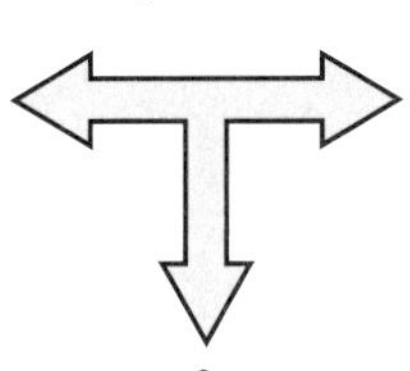

2

3

3 The kite ABCD is shown on the right. The diagonals AC and BD intersect at X.

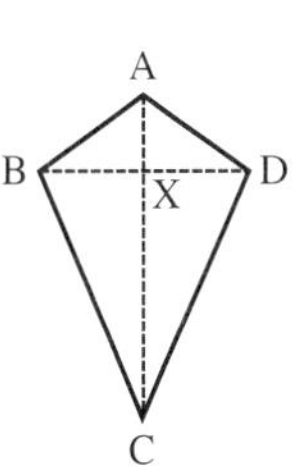

Which of the following statements are true?

a Triangle ABC is congruent to triangle ACD.

b Triangle ABD is congruent to triangle BCD.

c Triangle XBC is congruent to triangle XCD.

HOMEWORK 18B

1 On squared paper, show how each of these shapes tessellate. You should draw at least six shapes.

a

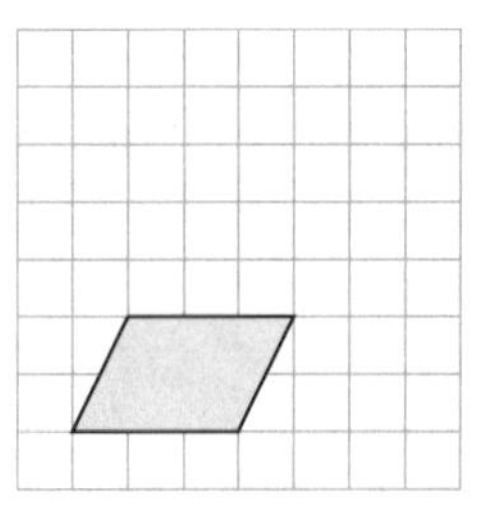

b

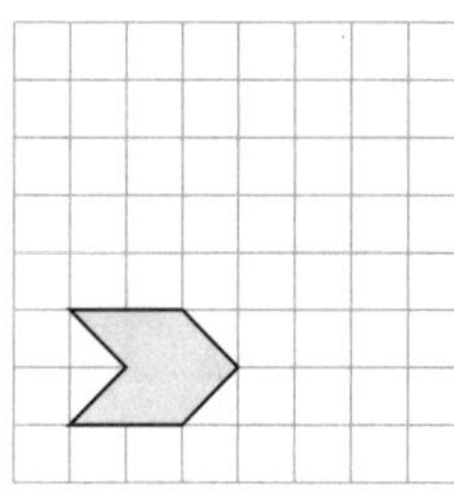

c

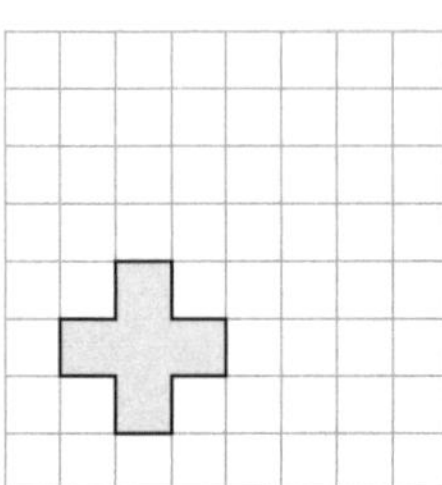

d

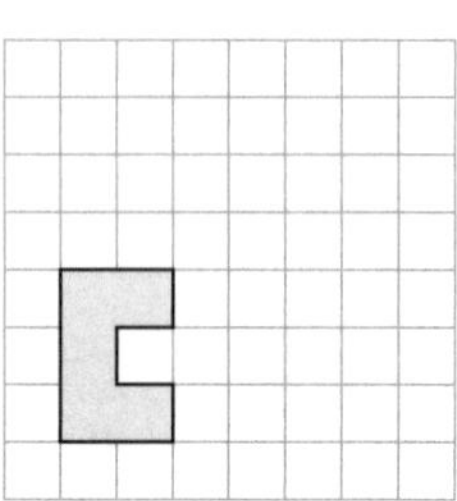

2 Use isometric paper to show how a regular hexagon tessellates.

HOMEWORK 18C

1 Copy each of these shapes on squared paper and draw its image by using the given translation.

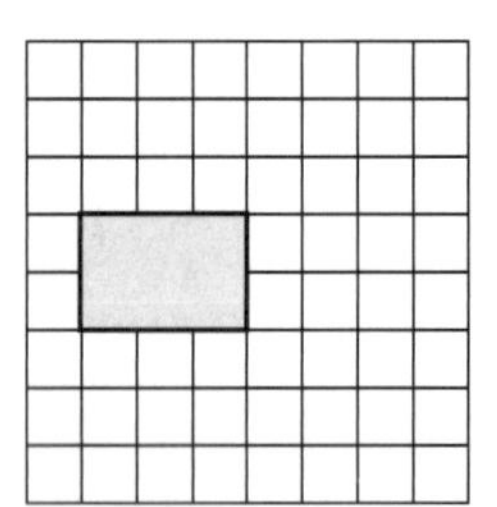

a 4 squares right

b 4 squares up

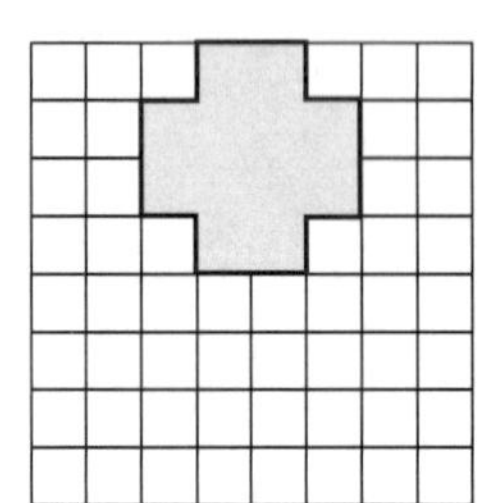

c 4 squares down

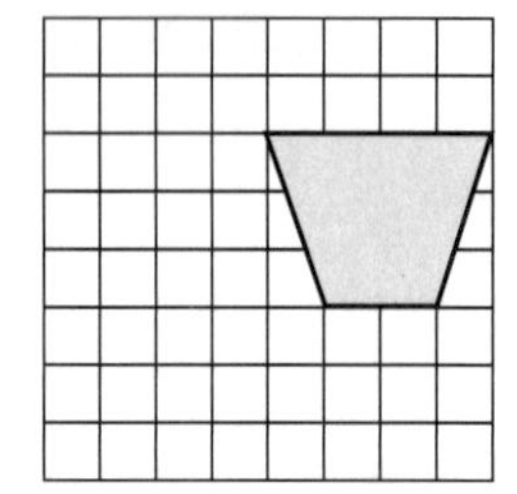

d 4 squares left

2 Copy each of these shapes on squared paper and draw its image by using the given translation.

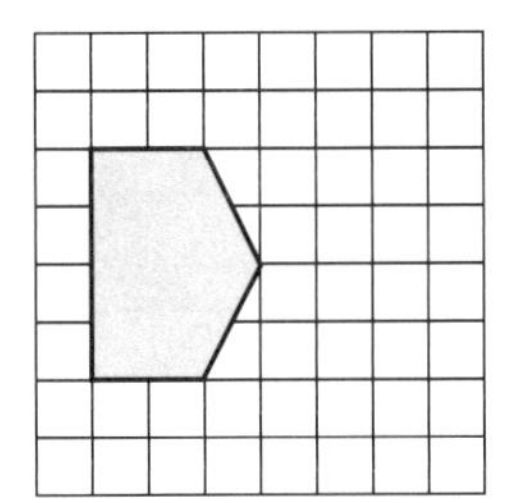

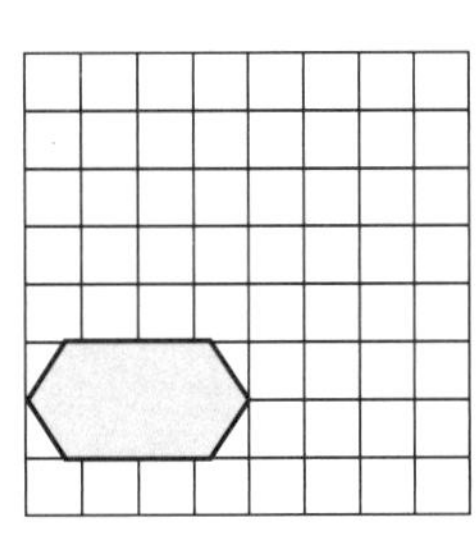

a 3 squares right and 2 squares down **b** 3 squares right and 4 squares up

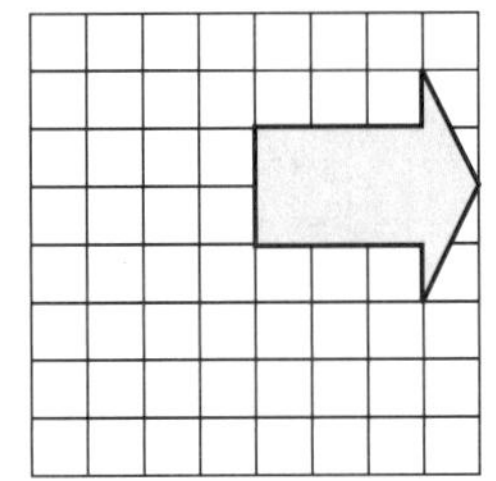

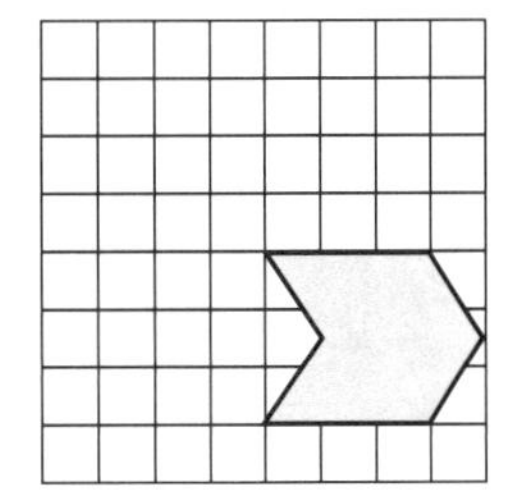

c 3 squares left and 3 squares down **d** 4 squares left and 1 square up

3 Describe these translations using vectors.

i A to B **ii** A to C **iii** A to D  **iv** B to A **v** B to C  **vi** B to D

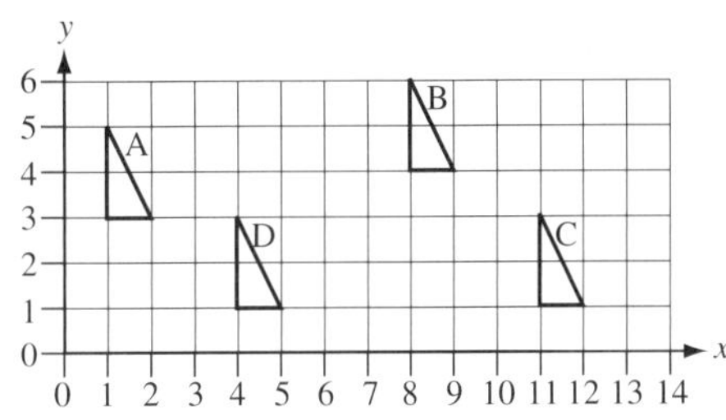

4 **a** On a grid showing values of x and y from 0 to 10, draw the triangle with co-ordinates A(4, 4), B(5, 7) and C(6, 5).

b Draw the image of ABC after a translation with vector $\binom{3}{2}$. Label this P.

c Draw the image of ABC after a translation with vector $\binom{4}{-3}$. Label this Q.

d Draw the image of ABC after a translation with vector $\binom{-4}{3}$. Label this R.

e Draw the image of ABC after a translation with vector $\binom{-3}{-2}$. Label this S.

HOMEWORK 18D

1 Copy each shape on squared paper and draw its image after a reflection in the given mirror line.

a

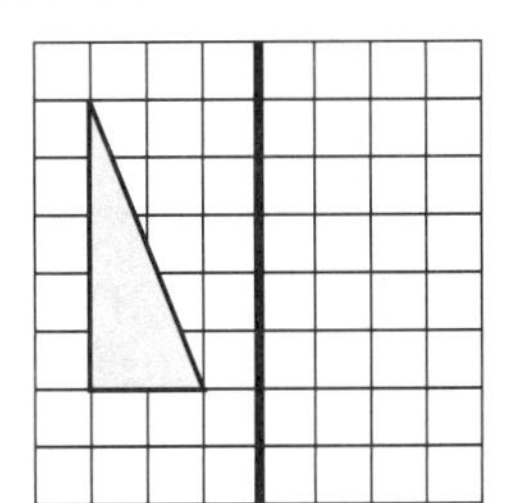

b

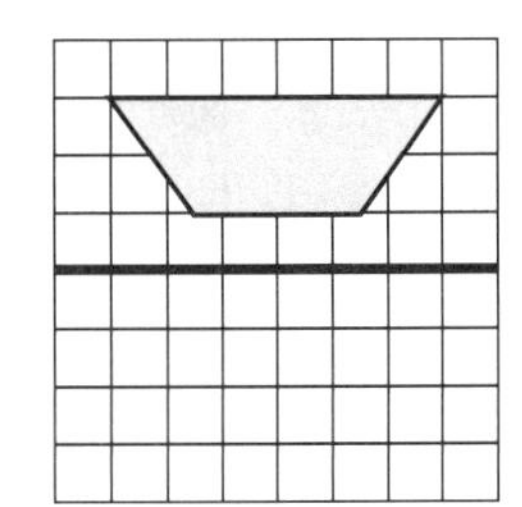

c

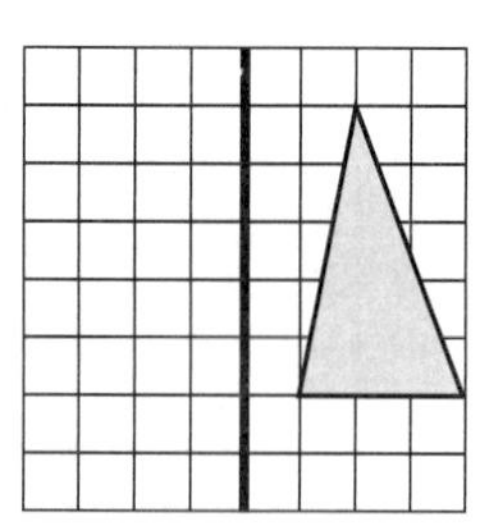

d

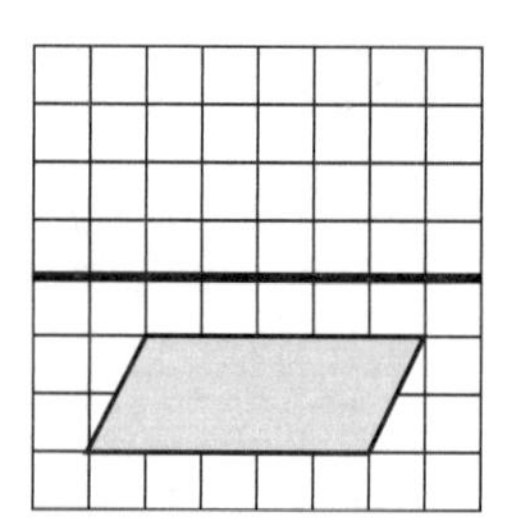

2 Draw each of these figures on squared paper and then draw the reflection of the figure in the mirror line.

a

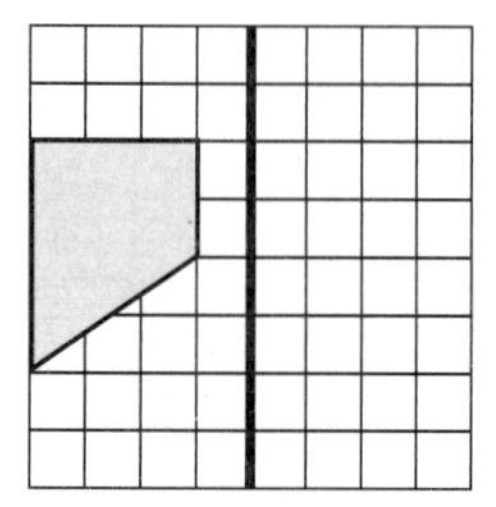

b

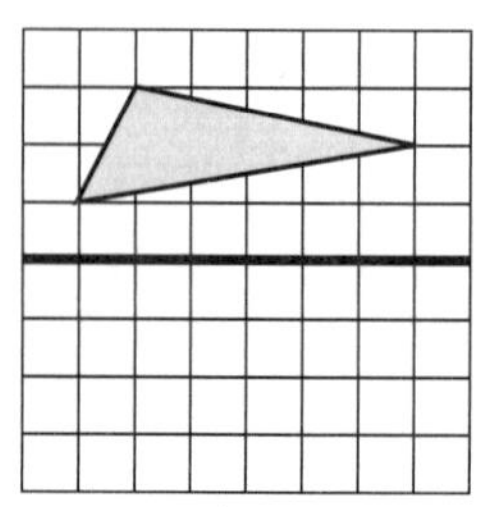

c

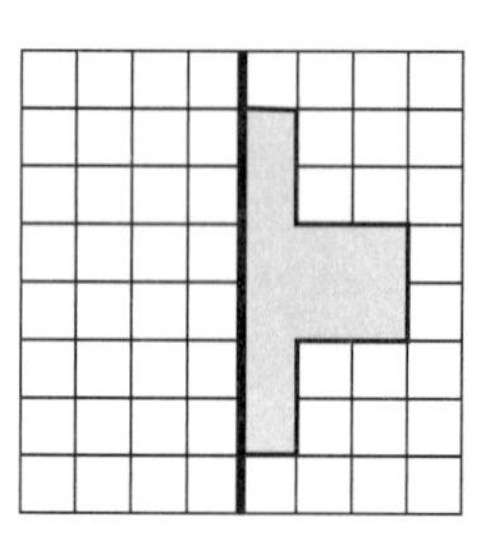

d

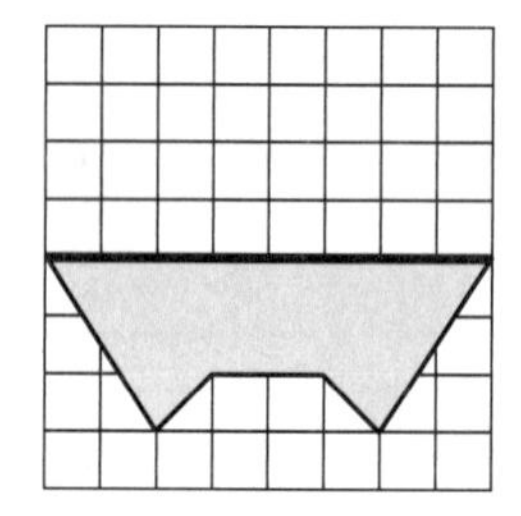

★3 Copy this diagram on squared paper.

a Reflect the triangle ABC in the x-axis. Label the image R.

b Reflect the triangle ABC in the y-axis. Label the image S.

c What special name is given to figures that are exactly the same shape and size?

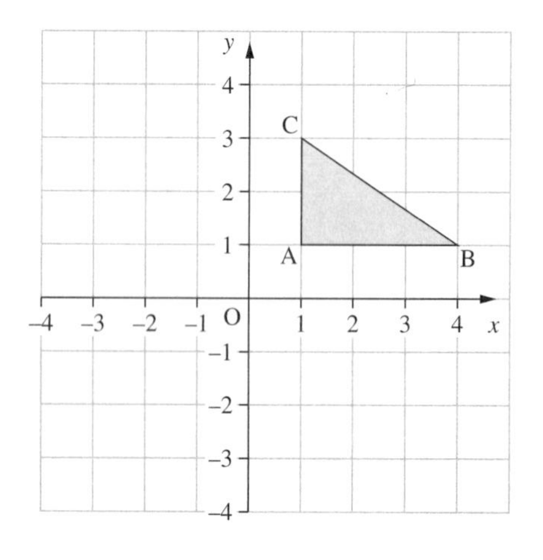

★4 **a** Draw a pair of axes, x-axis from –5 to 5, y-axis from –5 to 5.

b Draw the triangle with co-ordinates A(2, 2), B(3, 4), C(2, 4).

c Reflect the triangle ABC in the line $y = x$. Label the image P.

d Reflect the triangle P in the line $y = -x$. Label the image Q.

e Reflect triangle Q in the line $y = x$, label it R.

f Describe the reflection that will move triangle ABC to triangle R.

1 Copy each of these diagrams on squared paper. Draw its image using the given rotation about the centre of rotation A. Using tracing paper may help.

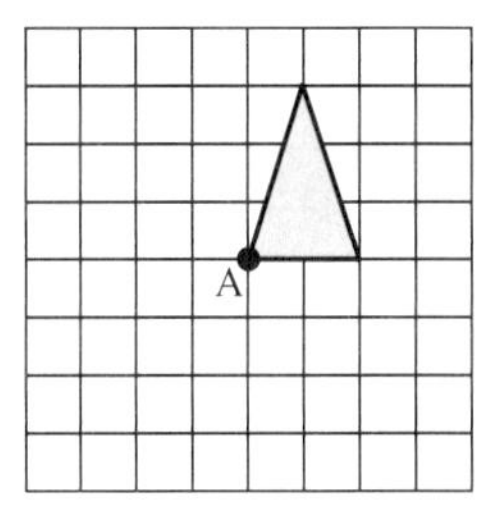

a $\frac{1}{2}$ turn

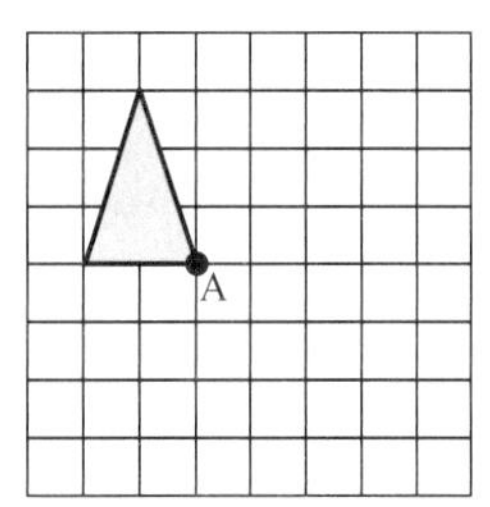

b $\frac{1}{4}$ turn clockwise

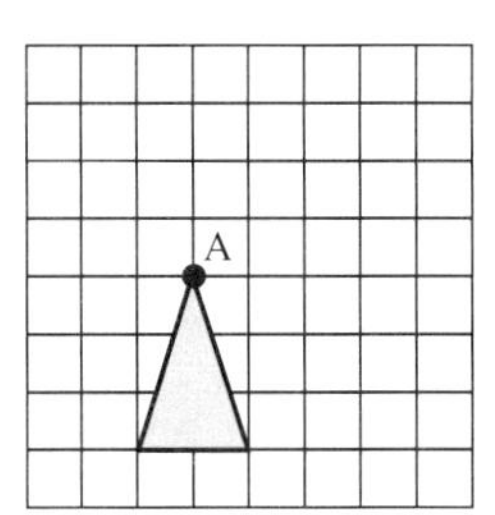

c $\frac{1}{4}$ turn anticlockwise

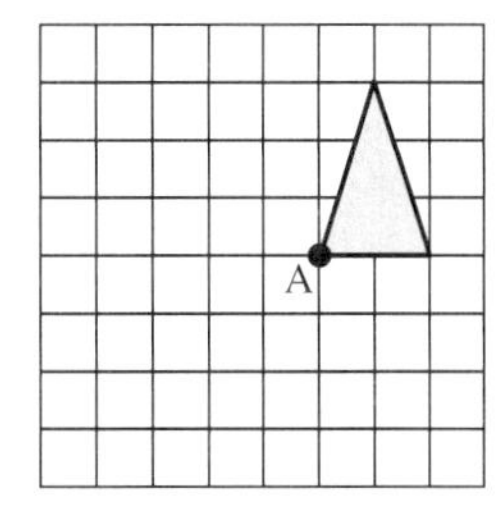

d $\frac{3}{4}$ turn clockwise

2 Copy each of these flags on squared paper. Draw its image using the given rotation about the centre of rotation A. Using tracing paper may help.

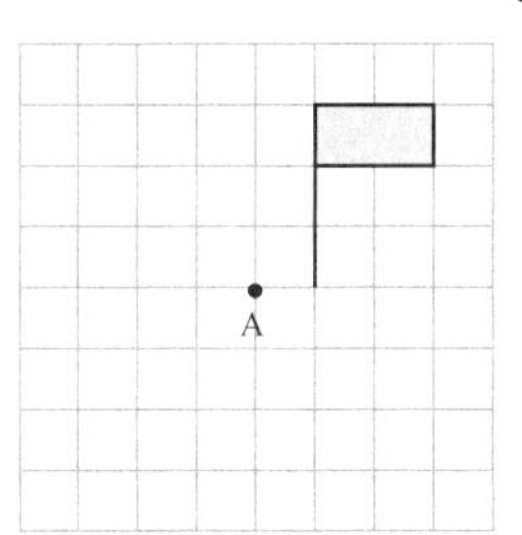

a 180° turn

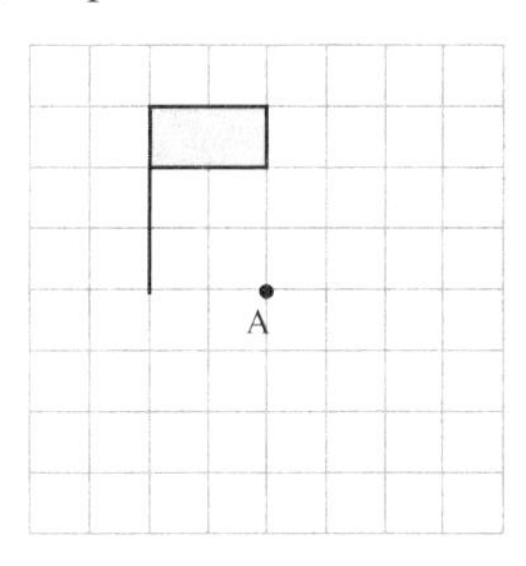

b 90° turn clockwise

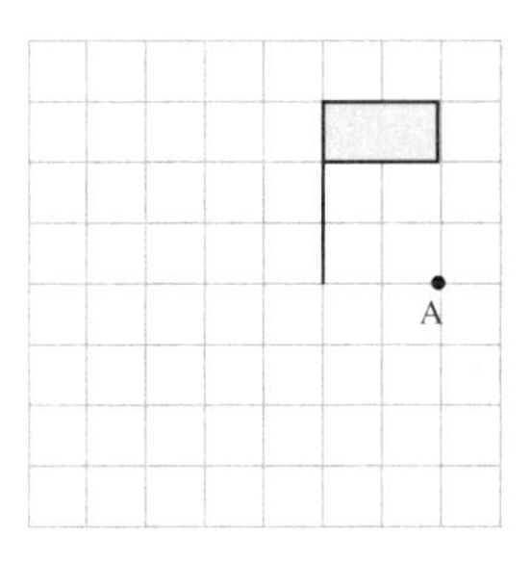

c 90° turn anticlockwise

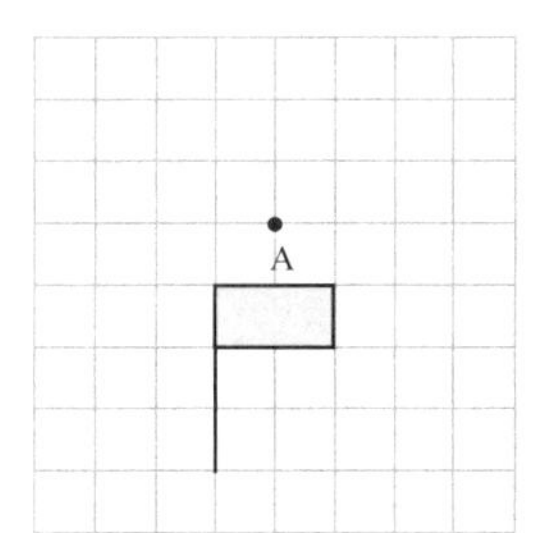

d 270° turn clockwise

★**3** Copy this T-shape on squared paper.

a Rotate the shape 90° clockwise about the origin O. Label the image P.

b Rotate the shape 180° about the origin O. Label the image Q.

c Rotate the shape 270° clockwise about the origin O. Label the image R.

d What rotation takes R back to the original shape?

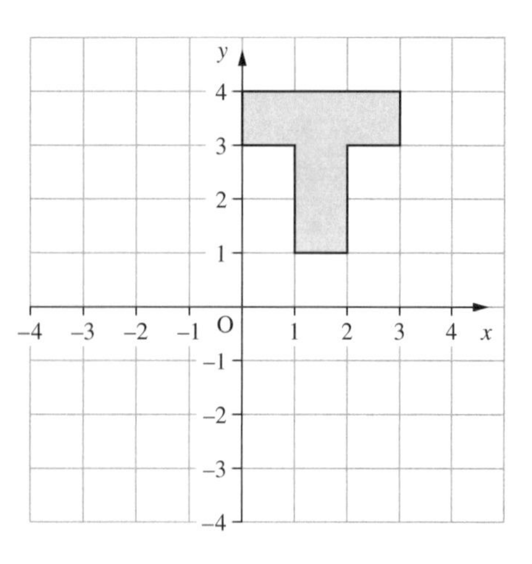

★**4** Copy the diagram and rotate the given triangle by:

a 90° clockwise about (0, 0)

b 180° about (0, –2)

c 90° anticlockwise about (–1, –1)

d 180° about (0, 0).

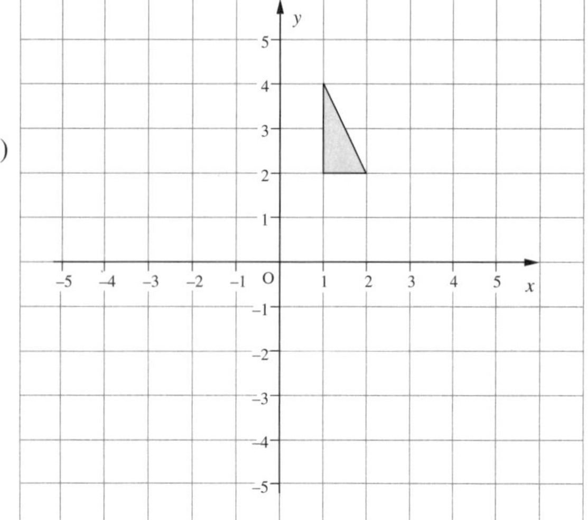

HOMEWORK 18F

1 Copy each of these figures on squared paper with its centre of enlargement A. Then enlarge it by the given scale factor using the ray method.

a

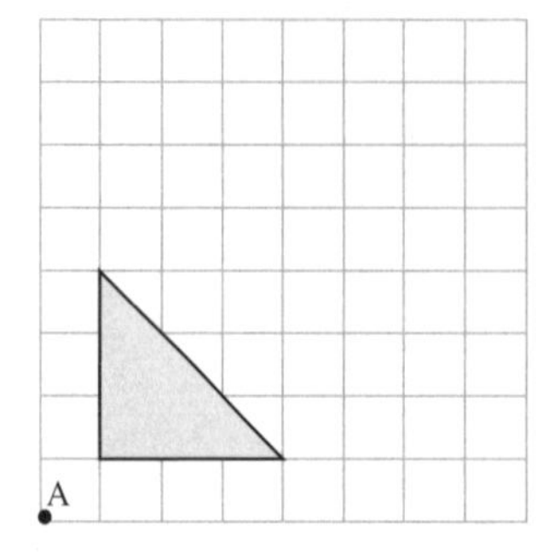

Scale factor 2

b

A

Scale factor 3

2 Copy each of these diagrams on squared paper and enlarge it by scale factor 2 using the origin as the centre of enlargement.

a

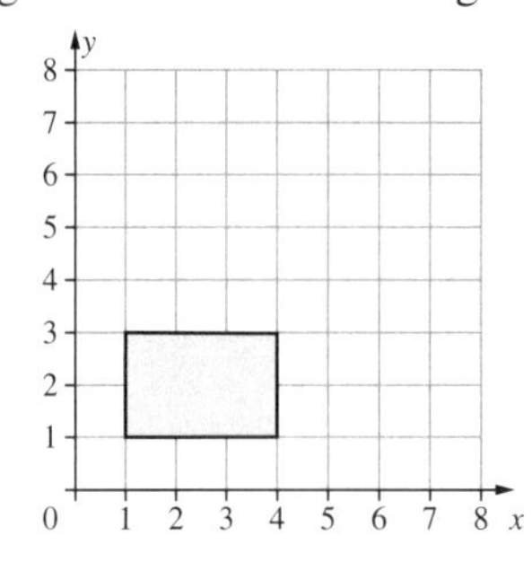

b

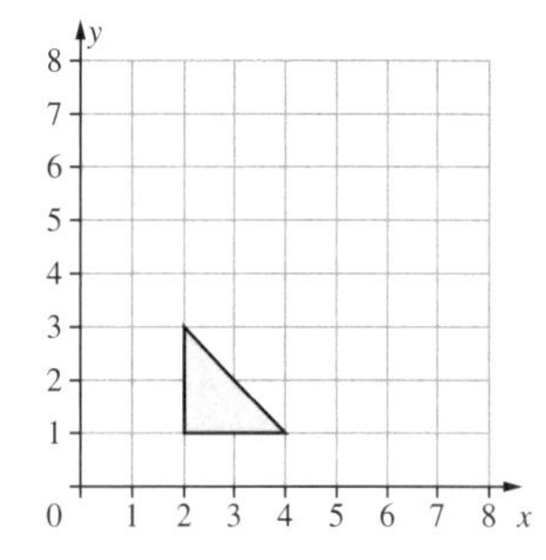

c

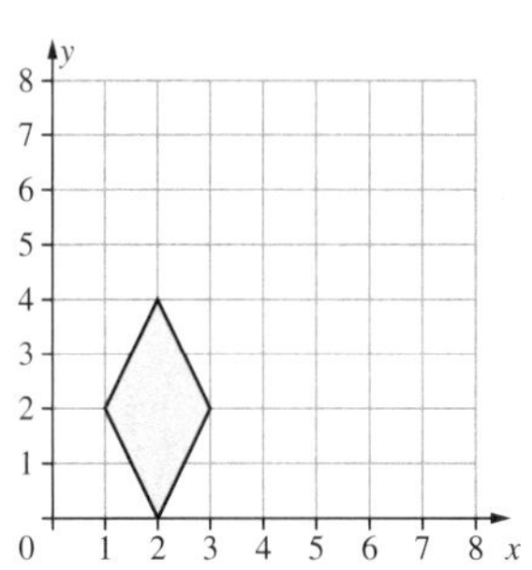

d

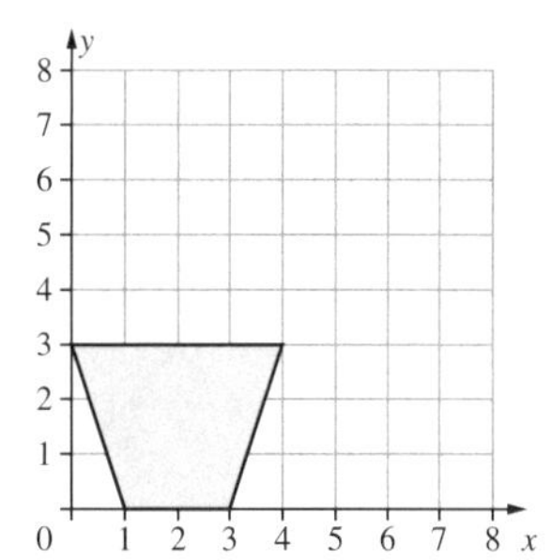

3 Copy each figure below with its centre of enlargement, leaving plenty of space for the enlargement. Then enlarge them by the given scale factor using the counting squares method.

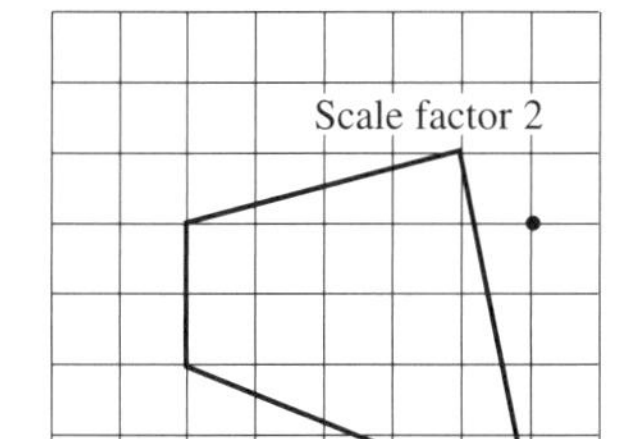

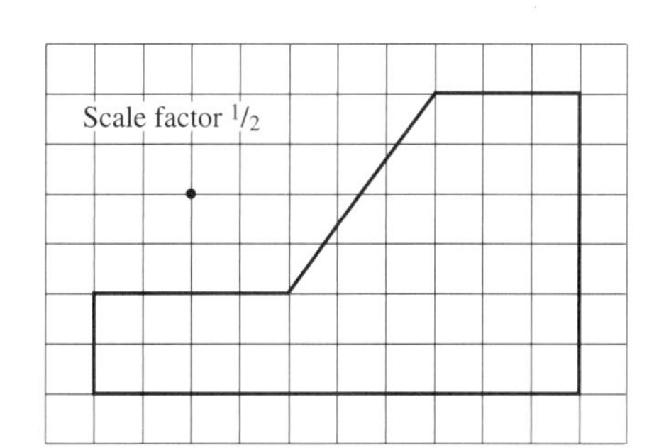

19 Constructions

HOMEWORK 19A

1 Accurately draw each of the following triangles.

a

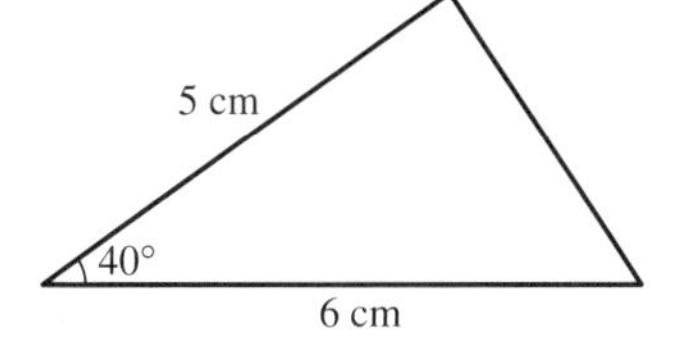

b

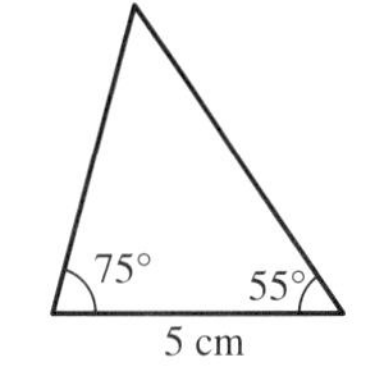

c

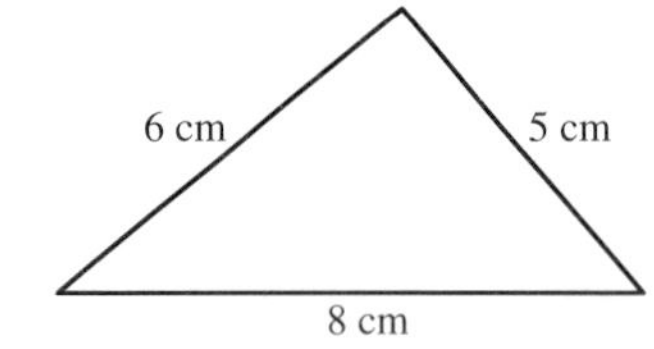

d

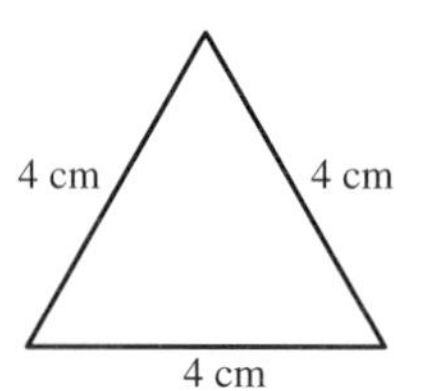

e

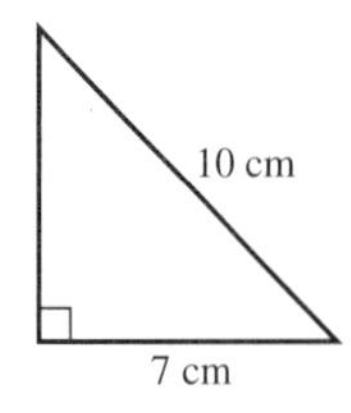

2 Draw a triangle ABC with AB = 6 cm, ∠A = 60° and ∠B = 50°.

3

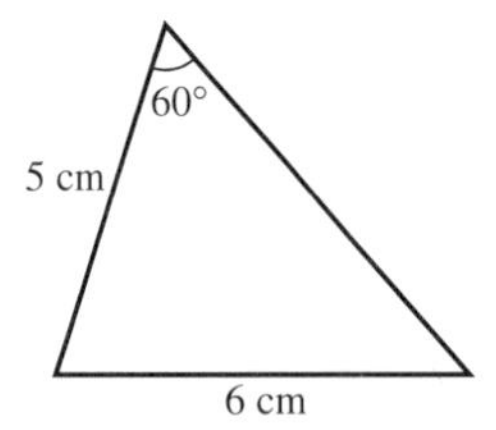

Explain why you can or cannot draw this triangle accurately.

4 **a** Accurately draw the shape on the right.

b What is the name of the shape you have drawn?

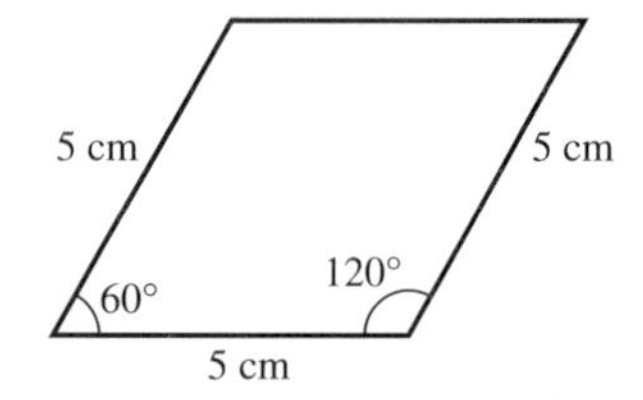

HOMEWORK 19B

1 Draw a line 8 cm long. Bisect it with a pair of compasses. Check your accuracy by seeing if each half is 4 cm.

2 **a** Draw any triangle.

b On each side construct the line bisector. All your line bisectors should intersect at the same point.

c See if you can use this point as the centre of a circle that fits perfectly inside the triangle.

3 **a** Draw a circle with a radius of about 4 cm.

b Draw a quadrilateral such that the vertices (corners) of the quadrilateral are on the circumference of the circle.

c Bisect two of the sides of the quadrilateral. Your bisectors should meet at the centre of the circle.

4 **a** Draw any angle.

b Construct the angle bisector.

c Check how accurate you have been by measuring each half.

★5 The diagram shows a park with two ice-cream sellers A and B. People always go to the ice-cream seller nearest to them. Shade the region of the park from which people go to ice-cream seller B.

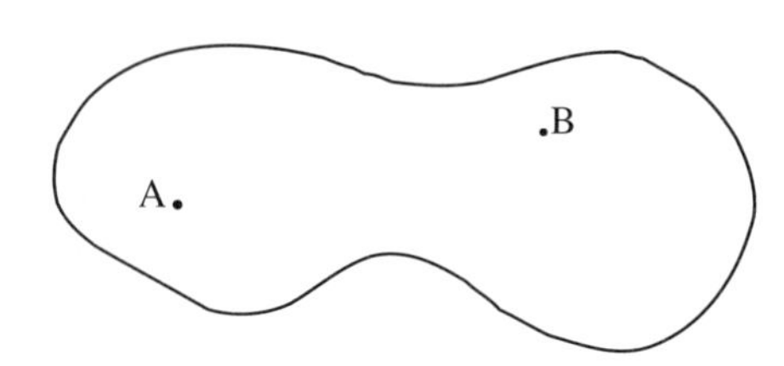

HOMEWORK 19C

1 A is a fixed point. Sketch the locus of the point P when AP > 3 cm and AP < 6 cm.

2 A and B are two fixed points 4 cm apart. Sketch the locus of the point P for the following situations:

a AP < BP **b** P is always within 3 cm of A and within 2 cm of B.

3 A fly is tethered by a length of spider's web that is 1 m long. Describe the locus that the fly can still buzz about in.

4 ABC is an equilateral triangle of side 4 cm. In each of the following loci, the point P moves only inside the triangle. Sketch the locus in each case.

a AP = BP **b** AP < BP

c CP < 2 cm **d** CP > 3 cm and BP > 3 cm

5 A wheel rolls around the inside of a square. Sketch the locus of the centre of the wheel.

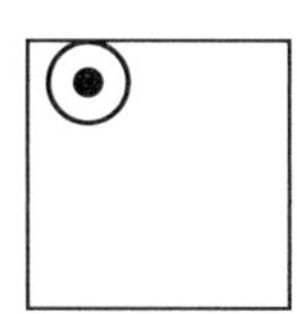

6 The same wheel rolls around the outside of the square. Sketch the locus of the centre of the wheel.

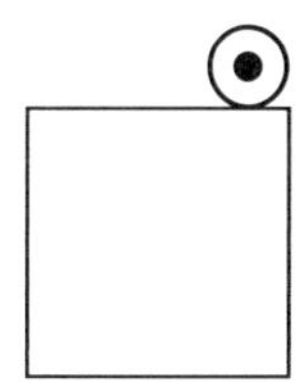

★7 Two ships A and B, which are 7 km apart, both hear a distress signal from a fishing boat. The fishing boat is less than 4 km from ship A and is less than 4.5 km from ship B. A helicopter pilot sees that the fishing boat is nearer to ship A than to ship B. Use accurate construction to show the region which contains the fishing boat. Shade this region.

HOMEWORK 19D

For Questions 1 to 3, you should start by sketching the picture given in each question on a 6 × 6 grid, each square of which is 1 cm by 1 cm. The scale for each question is given.

1 A goat is tethered by a rope, 10 m long, and a stake that is 2 m from each side of a field. What is the locus of the area that the goat can graze? Use a scale of 1 cm : 2 m.

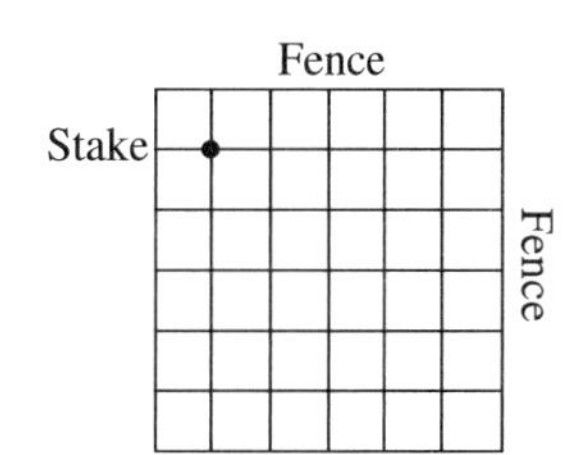

2 A cow is tethered to a rail at the top of a fence 4 m long. The rope is 4 m long. Sketch the area that the cow can graze. Use a scale of 1 cm : 2 m.

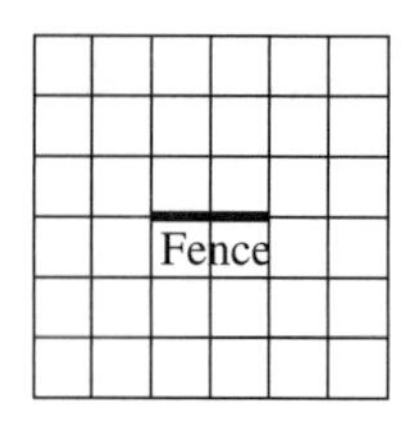

3 A horse is tethered to a corner of a shed, 3 m by 1 m. The rope is 4 m long. Sketch the area that the horse can graze. Use a scale of 1 cm : 1 m.

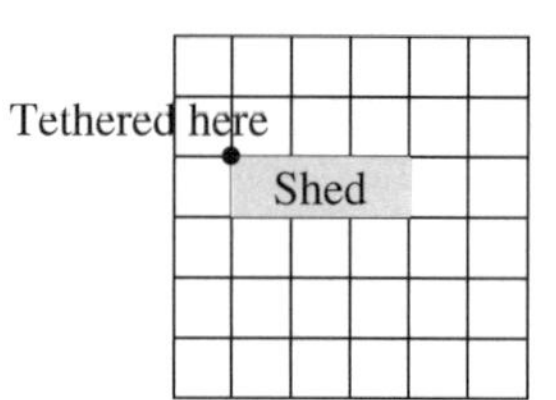

For Questions 4 to 6, you should use a copy of the map on page 102. For each question, trace the map and mark on those points that are relevant to that question.

4 A radio station broadcasts from Birmingham with a range that is just far enough to reach York. Another radio station broadcasts from Glasgow with a range that is just far enough to reach Newcastle.

a Sketch the area to which each station can broadcast.

b Will the Birmingham station broadcast as far as Norwich?

c Will the two stations interfere with each other?

5 An air traffic control centre is to be built in Newcastle. If it has a range of 200 km, will it cover all the area of Britain north of Sheffield and south of Glasgow?

6 A radio transmitter is to be built so that it is the same distance from Exeter, Norwich and Newcastle.

a Draw the perpendicular bisectors of the lines joining these three places to find where the station is to be built.

b Birmingham has so many radio stations that it cannot have another one within 50 km. Can the transmitter be built?

20 Units

HOMEWORK 20A

Decide in which metric unit you would most likely measure each of the following amounts.

1 The height of your best friend.

2 The distance from School to your home.

3 The thickness of a CD.

4 The weight of your maths teacher.

5 The amount of water in a lake.

6 The weight of a slice of bread.

7 The length of a double decker bus.

8 The weight of a kitten.

Estimate the approximate metric length, weight or capacity of each of the following.

9 This book (both length and weight).

10 The length of the road you live on. (You do not need to walk along it all.)

11 The capacity of a bottle of wine (metric measure).

12 A door (length, width and weight).

13 The diameter of a £1 coin, and its weight.

14 The distance from your school to The Houses of Parliamant (London).

HOMEWORK 20B

Length 10 mm = 1 cm, 1000 mm = 100 cm = 1 m, 1000 m = 1 km
Weight 1000 gm = 1 kg , 1000 kg = 1 t
Capacity 10 m*l* = 1 c*l*, 1000 m*l* = 100 c*l* = 1 litre
Volume 1000 litres = 1 m^3, 1 m*l* = 1 cl^3

Fill in the gaps using the information above.

1 155 cm = ... m

2 95 mm = ... cm

3 780 mm = ... m

4 3100 m = ... km

5 310 cm = ... m

6 3050 mm = ... m

7 156 mm = ... cm

8 2180 m = ... km

9 1070 mm = ... m

10 1324 cm = ... m

11 175 m = ... km

12 83 mm = ... m

13 620 mm = ... cm

14 2130 cm = ... m

15 5120 m = ... km

16 8150 g = ... kg

17 2300 kg = ... t

18 32 m*l* = ... c*l*

19 1360 m*l* = ... *l*

20 580 c*l* = ... *l*

21 950 kg = ... t

22 120 g = ... kg

23 150 m*l* = ... *l*

24 350 c*l* = ... *l*

25 540 m*l* = ... c*l*

26 2060 kg = ... t

27 7500 m*l* = ... *l*

28 3800 g = ... kg

29 605 c*l* = ... *l*

30 15 m*l* = ... *l*

31 6300 *l* = ... m^3

32 45 m*l* = ... cm^3

33 2350 *l* = ... m^3

34 720 *l* = ... m^3

35 8.2 m = ... cm

36 71 km = ... m

37 8.6 m = ... mm

38 15.6 cm = ... mm

39 0.83 m = ... cm

40 5.15 km = ... m

41 1.85 cm = ... mm

42 2.75 m = ... cm

HOMEWORK 20C

Length	12 inches = 1 foot, 3 feet = 1 yard, 1760 yards = 1 mile
Weight	16 ounces = 1 pound, 14 pounds = 1 stone, 2240 pounds = 1 ton
Capacity	8 pints = 1 gallon

Fill in the gaps using the information above.

1 5 feet = ... inches

2 5 yards = ... feet

3 3 miles = ... yards

4 6 pounds = ... ounces

5 5 stones = ... pounds

6 2 tons = ... pounds

7 4 gallons = ... pints

8 7 feet = ... inches

9 2 yards = ... inches

10 11 yards = ... feet

11 5 pounds = ... ounces

12 72 inches = ... feet

13 6 stones = ... pounds

14 39 feet = ... yards

15 2 stones = ... ounces

16 4400 yards = ... miles

17 12 gallons = ... pints

18 2 miles = ... feet

19 84 inches = ... feet

20 105 pounds = ... stones

21 48 pints = ... gallons

22 48 ounces = ... pounds

23 21 feet = ... yards

24 22 400 pounds = ... tons

25 2 miles = ... inches

26 256 ounces = ... pounds

27 80 pints = ... gallons

28 280 pounds = ... stones

29 31 680 feet = ... miles

30 2 tons = ... ounces

HOMEWORK 20D

Length	**Weight**	**Capacity**
1 metre ≈ 39 inches	2.2 pounds ≈ 1 kilogram	1 litre ≈ 1.75 pints
1 foot ≈ 30 centimetres		1 gallon ≈ 4.5 litres
1 foot ≈ 12 inches		
5 miles ≈ 8 kilometres		

1 Change each of these weights in kilograms to pounds.
a 6 **b** 8 **c** 15 **d** 32 **e** 45

2 Change each of these weights in pounds to kilograms. (Give each answer to 1 dp.)
a 10 **b** 18 **c** 25 **d** 40 **e** 56

3 Change each of these capacities in litres to pints.
a 2 **b** 8 **c** 25 **d** 60 **e** 75

4 Change each of these capacities in pints to litres. (Give each answer to the nearest litre.)
a 7 **b** 20 **c** 35 **d** 42 **e** 100

5 Change each of these distances in miles to kilometres.

a 20 **b** 30 **c** 50 **d** 65 **e** 120

6 Change each of these distances in kilometres to miles.

a 16 **b** 24 **c** 40 **d** 72 **e** 300

7 Change each of these capacities in gallons to litres.

a 5 **b** 12 **c** 27 **d** 50 **e** 72

8 Change each of these capacities in litres to gallons.

a 18 **b** 45 **c** 72 **d** 270 **e** 900

9 Change each of these distances in metres to inches.

a 2 **b** 5 **c** 8 **d** 10 **e** 12

10 Change each of these distances in feet to centimetres.

a 3 **b** 5 **c** 7 **d** 10 **e** 30

11 Change each of these distances in inches to metres. (Give your answer to 1 dp.)

a 48 **b** 52 **c** 60 **d** 75 **e** 100

21 Surface area and volume of 3-D shapes

HOMEWORK 21A

Find the volume of each 3-D shape if the edge of each cube is 1 cm.

1

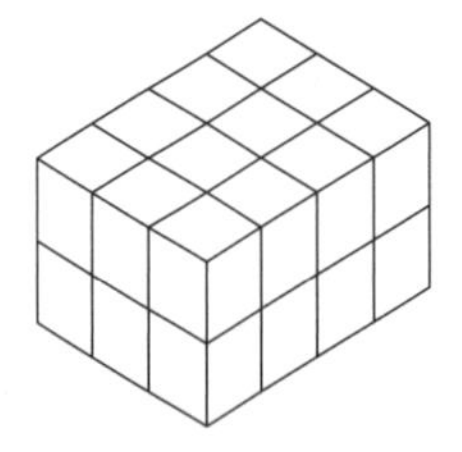

2

3

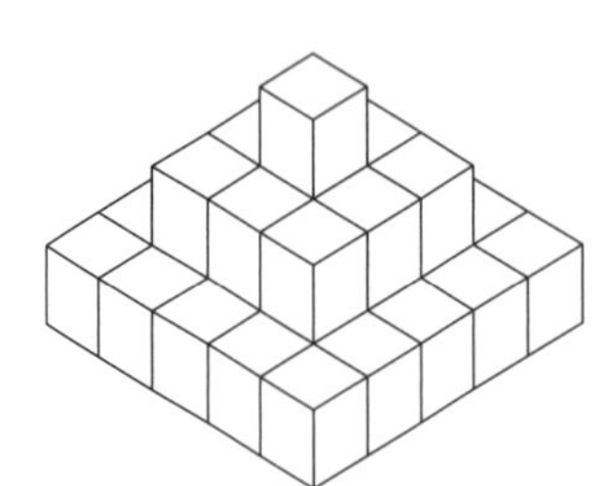

4

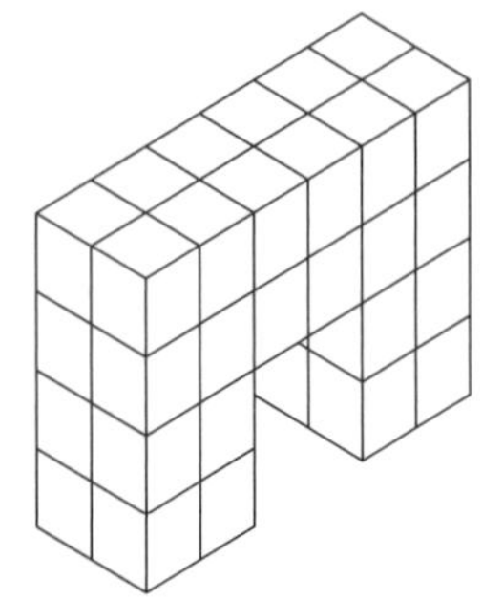

HOMEWORK 21B 

Example Calculate the volume and surface area of this cuboid.

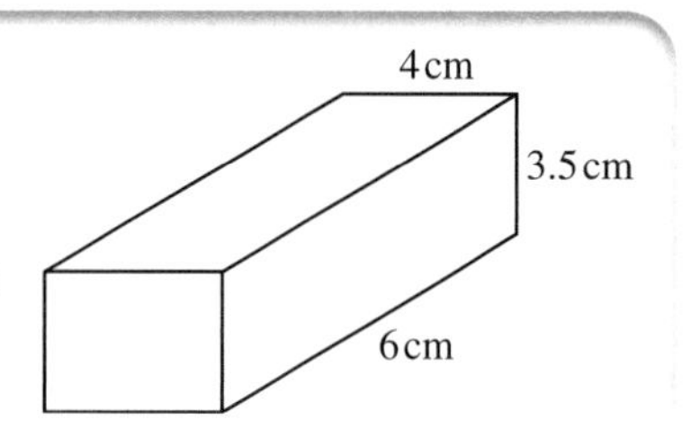

$$\text{Volume} = 6 \times 4 \times 3.5 = 84\text{ cm}^3$$

$$\text{Surface area} = (2 \times 6 \times 4) + (2 \times 3.5 \times 4) + (2 \times 3.5 \times 6) = 48 + 28 + 42 = 118\text{ cm}^2$$

1 Find **i** the volume and **ii** the surface area of each of these cuboids.

a

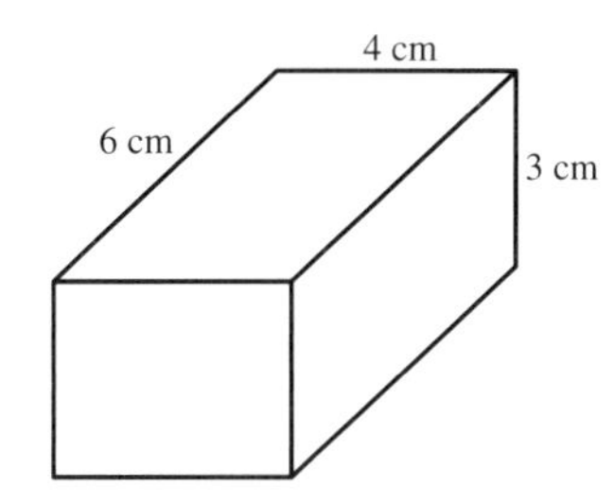

b

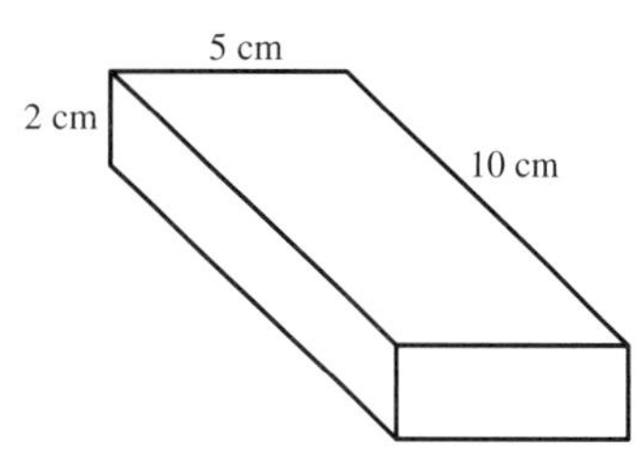

c

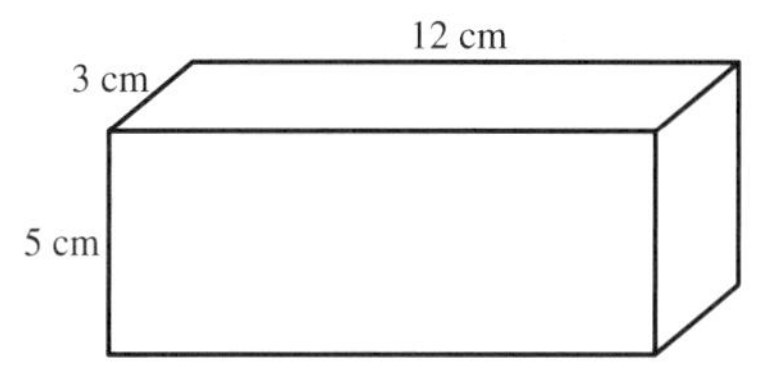

d

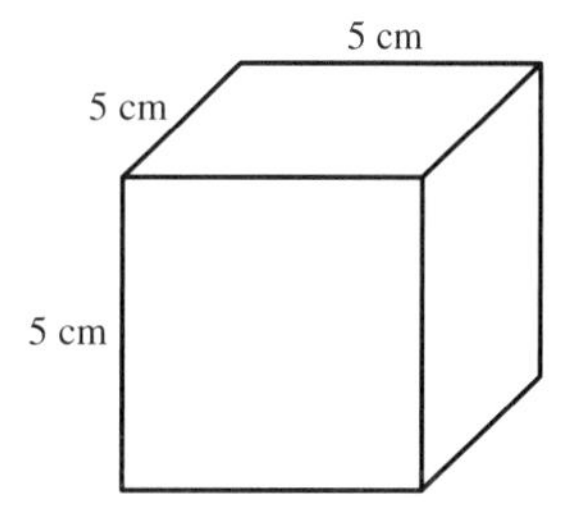

2 Copy and complete the table which shows the dimensions and volumes of four cuboids.

Length	Width	Height	Volume
4 cm	3 cm	2 cm	
	3 cm	3 cm	45 cm^3
8 cm		4 cm	160 cm^3
6 cm	6 cm		216 cm^3

3 Find the capacity (volume of a liquid or a gas) of a swimming pool whose dimensions are: length 12 m, width 5 m and depth 1.5 m.

4 Find the volume of the cuboid in each of the following cases.

a The area of the base is 20 cm^2 and the height is 3 cm.

b The base has one side 4 cm, the other side 1cm longer, and the height is 8 cm.

c The area of the top is 40 cm^2 and the depth is 3 cm.

HOMEWORK 21C 

The relationship between mass, density and volume can be expressed in three ways:

Mass = Density × Volume Density = Mass ÷ Volume Volume = Mass ÷ Density

1 Find the density of a piece of wood weighing 135 g and having a volume of 150 cm^3.

2 Calculate the density of a metal if 40 cm^3 of it weighs 2500 g.

3 Calculate the weight of a piece of plastic, 25 cm^3 in volume, if its density is 1.2 g/cm^3.

4 Calculate the volume of a piece of wood which weighs 350 g and has a density of 0.9 g/cm^3.

5 Find the weight of a marble statue, 540 cm^3 in volume, if the density of marble is 2.5 g/cm^3.

6 Calculate the volume of a liquid weighing 1 kg and having a density of 1.1 g/cm^3.

7 Find the density of the material of a stone which weighs 63 g and has a volume of 12 cm^3.

8 It is estimated that a huge rock balanced in the ceiling of a cave has a volume of 120 m^3. The density of the rock is 8.3 g/cm^3. What is the estimated weight of the rock?

9 A 1 kg bag of flour has a volume of about 900 cm^3. What is the density of flour in g/cm^3?

HOMEWORK 21D

1 For the prism below, calculate:

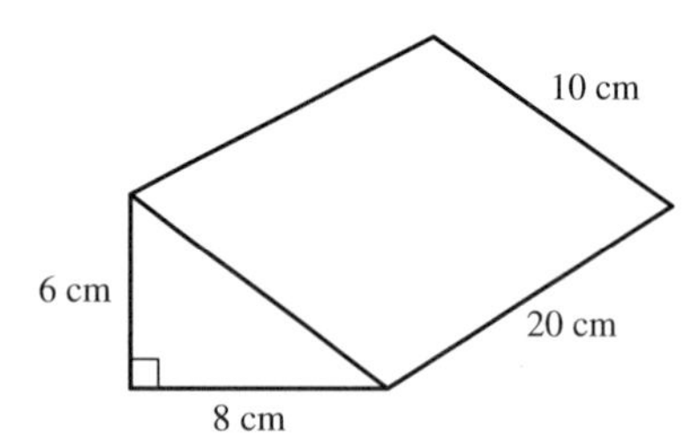

a its volume
b its total surface area.

2 For each prism shown, calculate **i** the area of the cross-section and **ii** the volume.

a

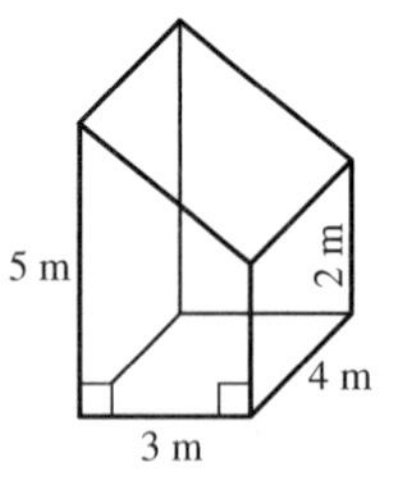

b

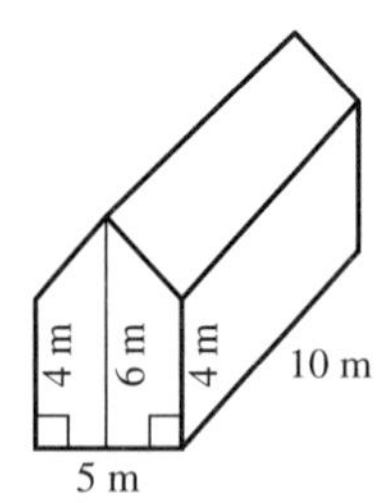

3 Calculate the weight of each prism.

a

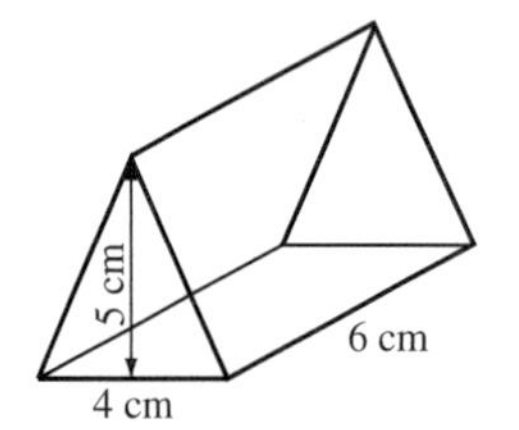

Density: 3.13 g/cm^3

b

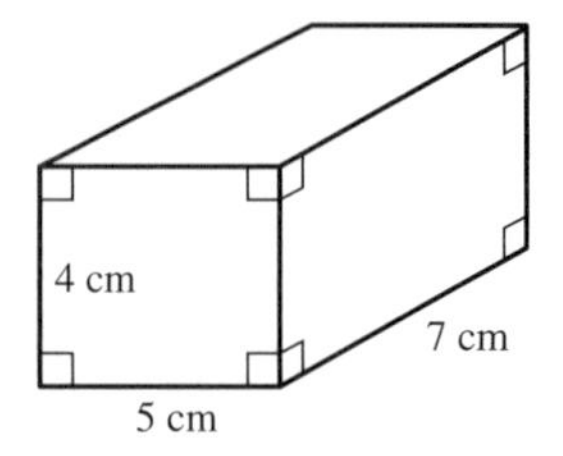

Density: 1.35g/cm^3

HOMEWORK 24E

Example Calculate the volume of a cylinder with a radius of 4 cm and a height of 10 cm.

$$\text{Volume} = \pi r^2 h = \pi \times 4^2 \times 10 = 502.7 \text{ cm}^3 \text{ (to 1 decimal place).}$$

1 Calculate the volume of each of these cylinders. Give your answers to one decimal place.

a Base radius 5 cm and a height of 7 cm.
b Base radius 10 cm and a height of 8 cm.
c Base diameter of 12 cm and a height of 20 cm.
d Base diameter of 9 cm and a height of 9 cm.

2 Find the volume of each of these cylinders. Give your answers to one decimal place.

a

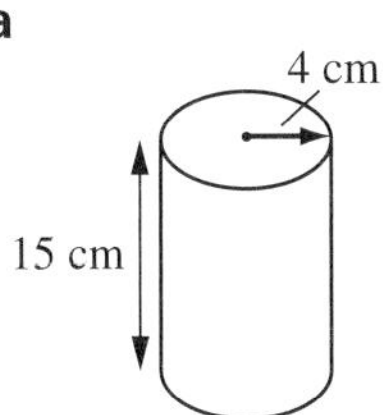

b

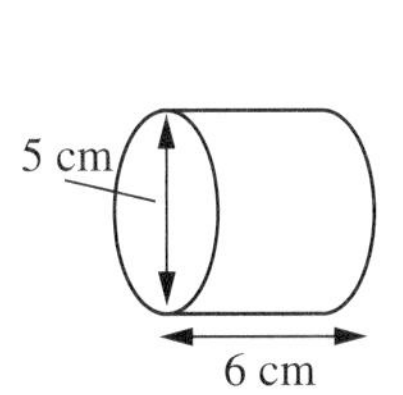

c

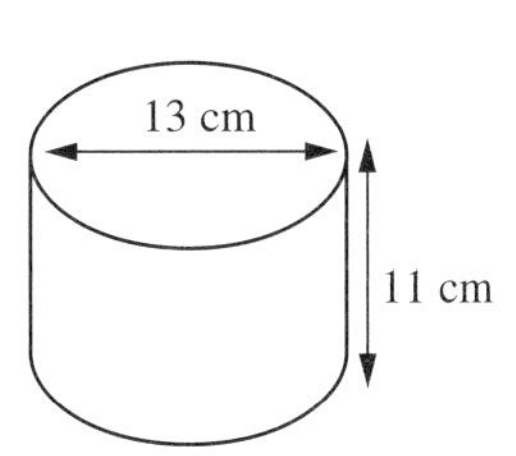

3 What is the weight of a solid iron bar 40 cm long with a radius of 2 cm? The density of iron is 8 grams per cm^3. Give your answer in kilograms.

4 Give the answers to this question in terms of π.

a What is the volume of a cylinder with a radius of 4 cm and a height of 11 cm?
b What is the volume of a cylinder with a diameter of 16 cm and a height of 18 cm?

22 Pythagoras' theorem

HOMEWORK 22A

For any right-angled triangle

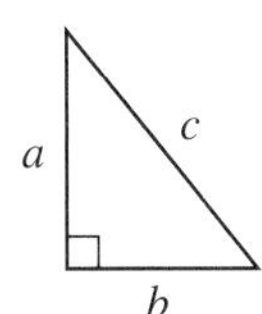

$$a^2 + b^2 = c^2$$

In each of the following right-angled triangles, calculate the length of the hypotenuse, x, giving your answers to one decimal place where necessary.

1

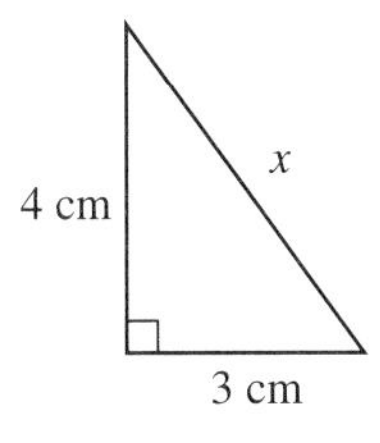

2

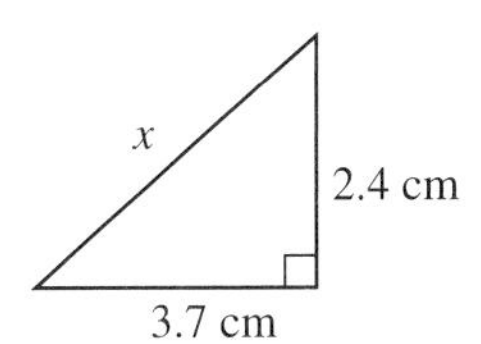

3

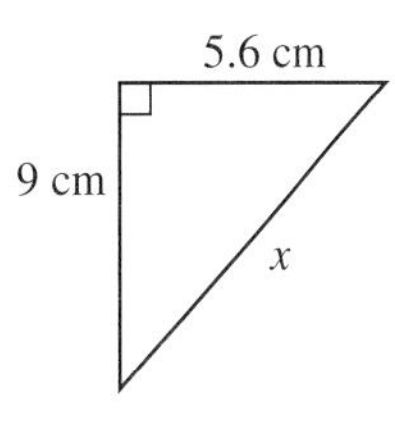

4

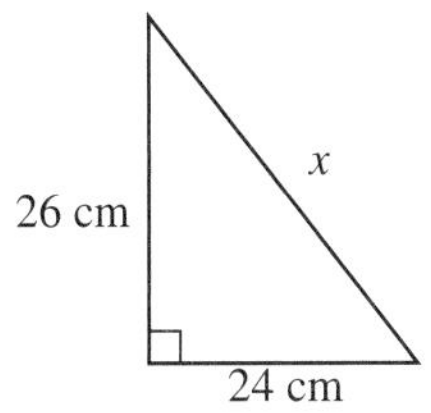

5

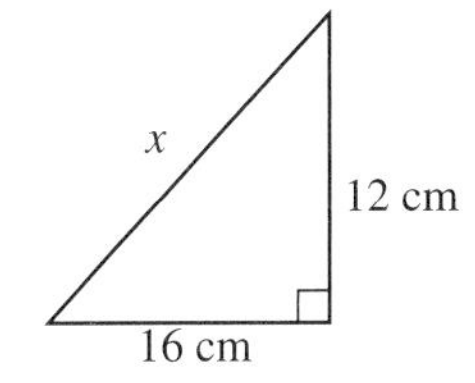

6

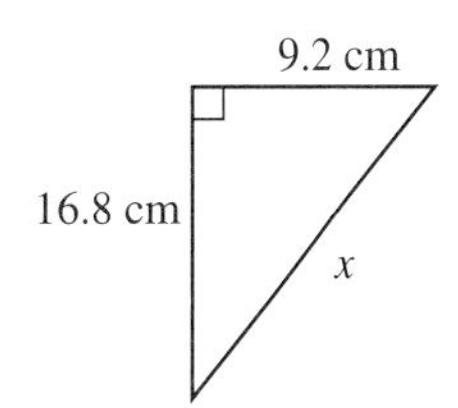

HOMEWORK 22B

For any right-angled triangle

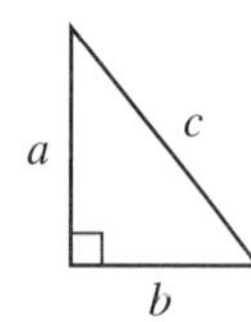

$a^2 = c^2 - b^2$ and $b^2 = c^2 - a^2$

1 In each of the following right-angled triangles, calculate the length of x, giving your answers to one decimal place where necessary.

a

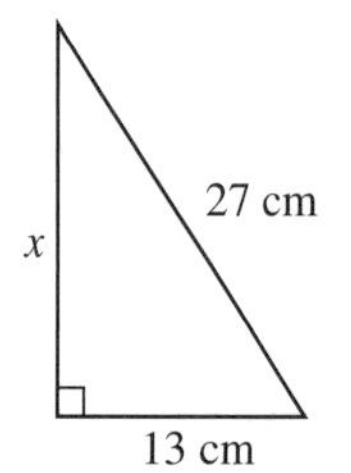

b

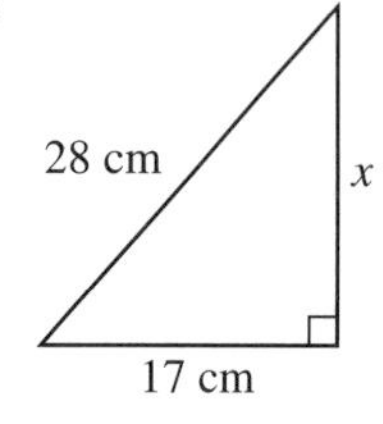

c

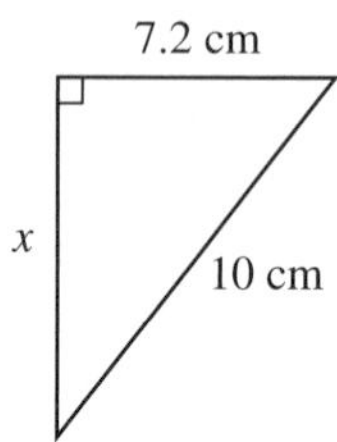

d

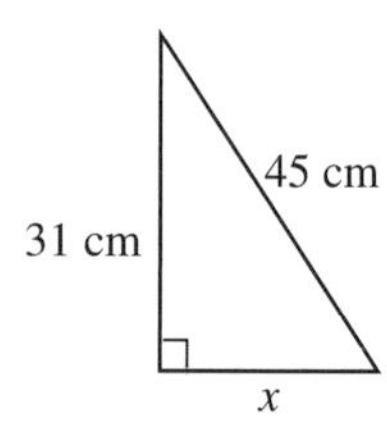

e

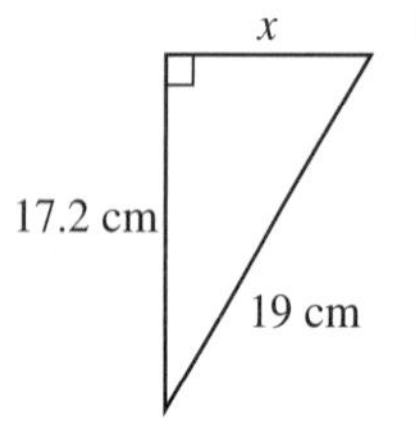

f

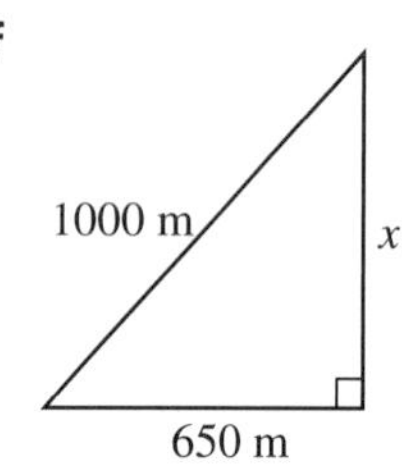

g

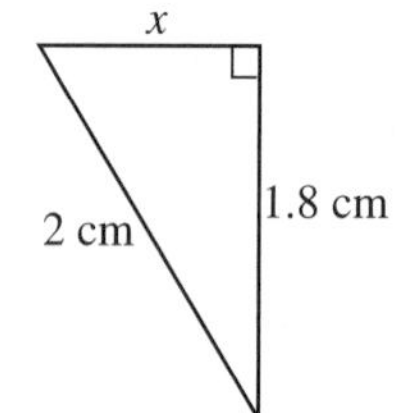

h

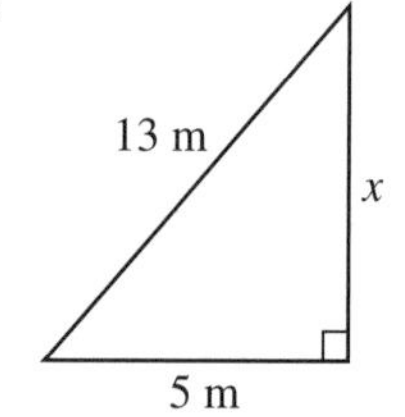

2 In each of the following right-angled triangles, calculate the length of x, giving your answers to one decimal place where necessary.

a

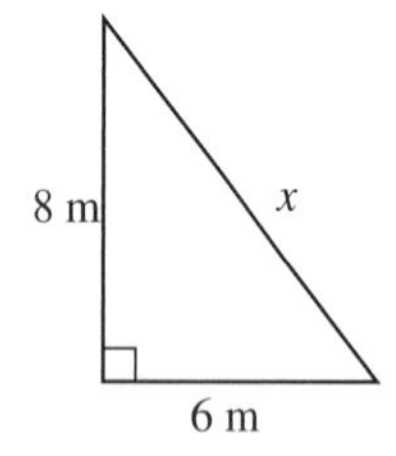

b

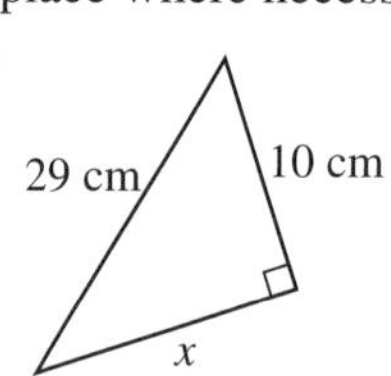

c

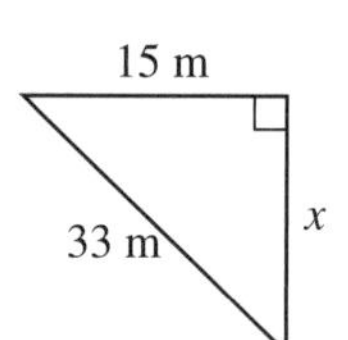

d

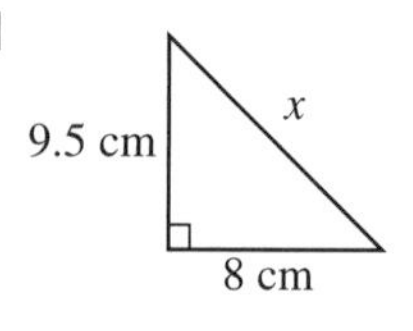

★**3** The diagram shows the end view of the framework for a sports arena stand. Calculate the length AB.

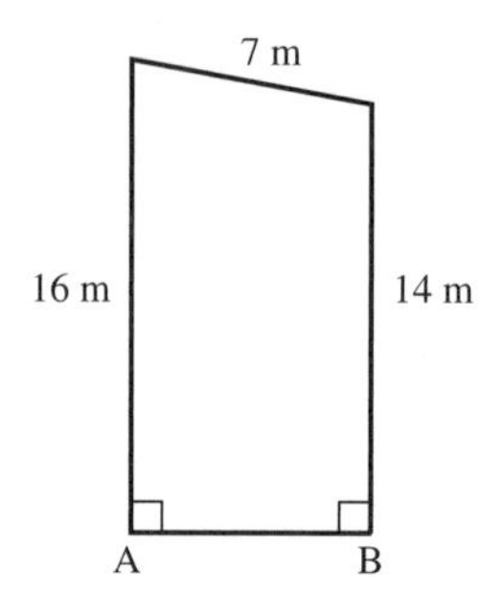

HOMEWORK 22C

1 A ladder, 15 metres long, leans against a wall. The ladder reaches 12 metres up the wall. How far away from the foot of the wall is the foot of the ladder? Give your answer to one decimal place.

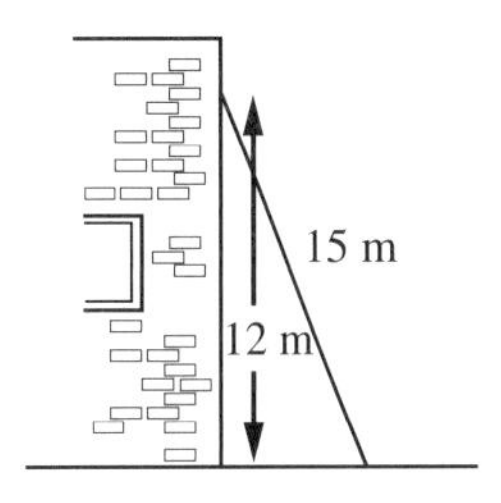

2 A rectangle is 3 metres long and 1.2 m wide. How long is the diagonal? Give your answer to one decimal place.

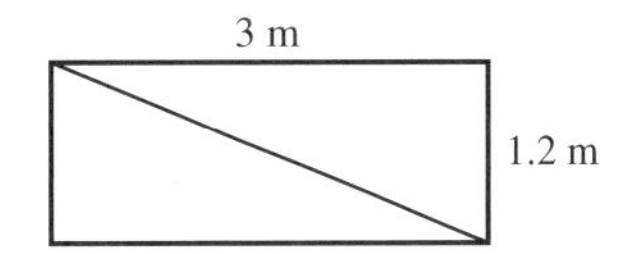

3 How long is the diagonal of a square with a side of 10 metres? Give your answer to one decimal place.

4 A ship going from a port to a lighthouse steams 8 km east and 6 km north. How far is the lighthouse from the port?

5 At the moment, three towns, A, B and C, are joined by two roads, as in the diagram. The council wants to make a road which runs directly from A to C. How much distance will the new road save? Give your answer to one decimal place.

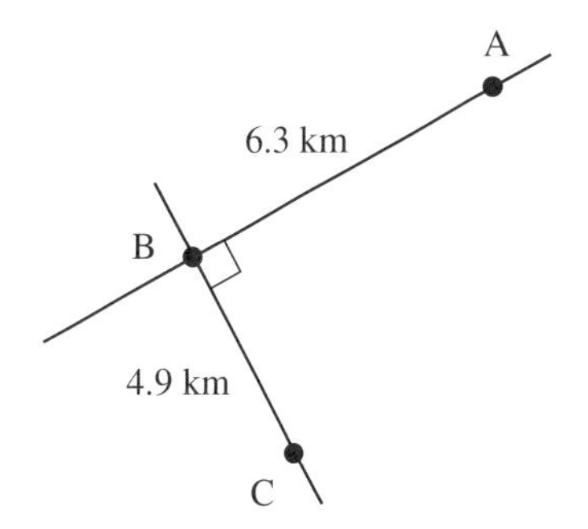

6 An 8 metre ladder is put up against a wall.

a How far up the wall will it reach when the foot of the ladder is 1 m away from the wall? Give your answer to one decimal place.

b When it reaches 7 m up the wall, how far is the foot of the ladder away from the wall? Give your answer to one decimal place.

7 How long is the line that joins the two co-ordinates A(1, 3) and B(2, 2)? Give your answer to one decimal place.

8 A rectangle is 4 cm long. The length of its diagonal is 5 cm. What is the area of the rectangle? Give your answer to one decimal place.

9 Is a triangle with sides 9 cm, 40 cm and 41 cm a right-angled triangle? Give your answer to one decimal place.

10 How long is the line that joins the two co-ordinates A(–3, –7), and B(4, 6)? Give your answer to one decimal place.

23 Basic algebra

HOMEWORK 23A

1 Write down the algebraic expression that says:

a 4 more than x **b** 7 less than x **c** k more than 3
d t less than 8 **e** x added to y **f** x multiplied by 4
g 5 multiplied by t **h** a multiplied by b **i** m divided by 2
j p divided by q.

2 Val is x years old. Dave is four years older than Val and Ella is five years younger than Val.

a How old is Dave? **b** How old is Ella?

3 A packet contains n sweets.

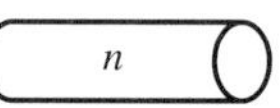

The total number of sweets here is $2n + 3$.

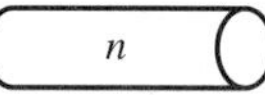
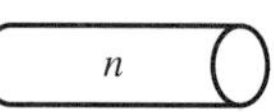

Write down an expression for the total number of sweets in the following.

a

b

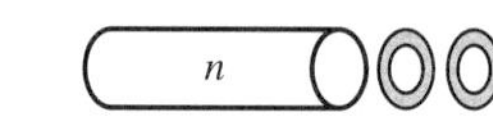

c

4 Sue has p pets.

- Frank has two more pets than Sue.
- Chloe has three less pets than Sue.
- Lizzie has twice as many pets as Sue.

How many pets does each person have?

5 **a** Tom has £20 and spends £16. How much does he have left?
b Sam has £10 and spends £a. How much does he have left?
c Ian has £b and spends £c. How much does he have left?

6 **a** How many days are there in 3 weeks?
b How many days are there in z weeks?

★**7** **a** Granny Parker divides £30 equally between her 3 grandchildren. How much does each receive?
b Granny Smith divides £r equally between her 4 grandchildren How much does each receive?
c Granny Thomas divides £p equally between her q grandchildren. How much does each receive?

HOMEWORK 23B

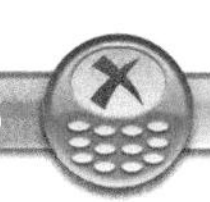

Evaluate these expressions, writing them as simply as possible.

1	$3 \times 4t$	**2**	$2 \times 5y$	**3**	$4y \times 2$	**4**	$3w \times 3$
5	$4t \times t$	**6**	$6b \times b$	**7**	$3w \times w$	**8**	$6y \times 2y$
9	$5p \times p$	**10**	$4t \times 32t$	**11**	$5m \times 4m$	**12**	$6t \times 4t$
13	$m \times 7t$	**14**	$5y \times w$	**15**	$8t \times q$	**16**	$n \times 69t$
17	$5 \times 6q$	**18**	$5f \times 2$	**19**	$6 \times 3k$	**20**	$5 \times 7r$
21	$t^2 \times t$	**22**	$p \times p^2$	**23**	$5m \times m^2$	**24**	$3t^2 \times t$
25	$4n \times 2n^2$	**26**	$5r^2 \times 4r$	**27**	$t^2 \times t^2$	**28**	$k^3 \times k^2$
29	$8n^2 \times 2n^3$	**30**	$4t^3 \times 3t^4$	**31**	$7a^4 \times 2a^3$	**32**	$k^5 \times 3k^2$
33	$-k^2 \times -k$	**34**	$-5y \times -2y$	**35**	$-3d^2 \times -6d$	**36**	$-2p^4 \times 6p^2$
37	$5mq \times q$	**38**	$4my \times 3m$	**39**	$4mt \times 3m$	**40**	$5qp \times 2qp$

HOMEWORK 23C

Example 1 $x + 3x + 2x - 4x = 2x$

Example 2 $2a + 3b + 5a + 2b - 4a - b = 3a + 4b$

Example 3 $2x^2 + 4x^2 - x^2 = 5x^2$

Example 4 $5x^2 + 3y - 3x^2 - 4y = 2x^2 - y$

1 Write each of these expressions in a shorter form.

a $a + a + a$ **b** $3b + 2b$ **c** $3c + c + 5c$
d $5d - d$ **e** $5e + 2e - 4e$ **f** $7f - 2f + 3f$
g $2g + 4g - 6g$ **h** $4h - 6h$ **i** $3i^2 + 2i^2$
j $5j^2 + j^2 - 2j^2$

2 Simplify each of the following expressions.

a $2y + 5x + y + 3x$ **b** $4m + 6p - 2m + 4p$ **c** $3x + 6 + 3x - 2$
d $7 - 5x - 2 + 8x$ **e** $5p + 2t + 3p - 2t$ **f** $4 + 2x + 4x - 6$
g $4p - 4 - 2p - 2$ **h** $4x + 3y + 2x - 5y$ **i** $4 + 3t + p - 6t + 3 + 5p$
j $4w - 3k - 2w - k + 4w$

3 Simplify each of the following expressions.

a $4x + 8 - 3x + 1$ **b** $7 - 3y - 4 + 5y$ **c** $5a + 3b - a - 5b$
d $5c - 8d - 3c + 4d$ **e** $7x + 3y + 3 + 5y - 6$ **f** $4a + 3b - 4a - b$

4 Simplify each of the following expressions.

a $3x^2 + 8 - 2x^2 - 3$ **b** $5a^2 + 3b - 4a^2 + 2b$ **c** $k + 3k^2 - 3k + 2k^2$
d $3c^2 + 4d - 3c^2 - 3d$ **e** $5x^2 + 3y^2 - 3x^2 + y^2$ **f** $4y^2 + 2z^2 - 6y^2 - 3z^2$

HOMEWORK 23D

Expand these expressions.

1	$3(4 + m)$	**2**	$6(3 + p)$	**3**	$4(4 - y)$	**4**	$3(6 + 7k)$
5	$4(3 - 5f)$	**6**	$2(4 - 23w)$	**7**	$7(g + h)$	**8**	$4(2k + 4m)$
9	$6(2d - n)$	**10**	$t(t + 5)$	**11**	$m(m + 4)$	**12**	$k(k - 2)$
13	$g(4g + 1)$	**14**	$y(3y - 21)$	**15**	$p(7 - 8p)$	**16**	$2m(m + 5)$
17	$3t(t - 2)$	**18**	$3k(5 - k)$	**19**	$2g(4g + 3)$	**20**	$4h(2h - 3)$
21	$2t(6 - 5t)$	**22**	$4d(3d + 5e)$	**23**	$3y(4y + 5k)$	**24**	$6m^2(3m - p)$
25	$y(y^2 + 7)$	**26**	$h(h^3 + 9)$	**27**	$k(k^2 - 4)$	**28**	$3t(t^2 + 3)$
29	$5h(h^3 - 2)$	**30**	$4g(g^3 - 3)$	**31**	$5m(2m^2 + m)$	**32**	$2d(4d^2 - d^3)$
33	$4w(3w^2 + t)$	**34**	$3a(5a^2 - b)$	**35**	$2p(7p^3 - 8m)$	**36**	$m^2(3 + 5m)$
37	$t^3(t + 3t)$	**38**	$g^2(4t - 3g^2)$	**39**	$2t^2(7t + m)$	**40**	$3h^2(4h + 5g)$

HOMEWORK 23E

1 Simplify these expressions.

a $5t + 4t$ **b** $4m + 3m$ **c** $6y + y$ **d** $2d + 3d + 5d$
e $7e - 5e$ **f** $6g - 3g$ **g** $3p - p$ **h** $5t - t$
i $t^2 + 4t^2$ **j** $5y^2 - 2y^2$ **k** $4ab + 3ab$ **l** $5a^2d - 4a^2d$

2 Expand and simplify.

a $3(2 + t) + 4(3 + t)$ **b** $6(2 + 3k) + 2(5 + 3k)$ **c** $5(2 + 4m) + 3(1 + 4m)$
d $3(4 + y) + 5(1 + 2y)$ **e** $5(2 + 3f) + 3(6 - f)$ **f** $7(2 + 5g) + 2(3 - g)$

3 Expand and simplify.

a $4(3 + 2h) - 2(5 + 3h)$ **b** $5(3g + 4) - 3(2g + 5)$ **c** $3(4y + 5) - 2(3y + 2)$
d $3(5t + 2) - 2(4t + 5)$ **e** $5(5k + 2) - 2(4k - 3)$ **f** $4(4e + 3) - 2(5e - 4)$

4 Expand and simplify.

a $m(5 + p) + p(2 + m)$ **b** $k(4 + h) + h(5 + 2k)$ **c** $t(1 + 2n) + n(3 + 5t)$
d $p(5q + 1) + q(3p + 5)$ **e** $2h(3 + 4j) + 3j(h + 4)$ **f** $3y(4t + 5) + 2t(1 + 4y)$

5 Expand and simplify.

a $t(2t + 5) + 2t(4 + t)$ **b** $3y(4 + 3y) + y(6y - 5)$ **c** $5w(3w + 2) + 4w(3 - w)$
d $4p(2p + 3) - 3p(2 - 3p)$ **e** $4m(m - 1) + 3m(4 - m)$ **f** $5d(3 - d) + d(2d - 1)$

6 Expand and simplify.

a $5a(3b + 2a) + a(2a^2 + 3c)$ **b** $4y(3w + y^2) + y(3y - 4t)$

HOMEWORK 23F

Factorise the following expressions.

1	$9m + 12t$	**2**	$9t + 6p$	**3**	$4m + 12k$	**4**	$4r + 6t$
5	$2mn + 3m$	**6**	$4g^2 + 3g$	**7**	$4w - 8t$	**8**	$10p - 6k$
9	$12h - 10k$	**10**	$4mp + 2mk$	**11**	$4bc + 6bk$	**12**	$8ab + 4ac$

13 $3y^2 + 4y$ **14** $5t^2 - 3t$ **15** $3d^2 - 2d$ **16** $6m^2 - 3mp$

17 $3p^2 + 9pt$ **18** $8pt + 12mp$ **19** $8ab - 6bc$ **20** $4a^2 - 8ab$

21 $8mt - 6pt$ **22** $20at^2 + 12at$ **23** $4b^2c - 10bc$ **24** $4abc + 6bed$

25 $6a^2 + 4a + 10$ **26** $12ab + 6bc + 9bd$ **27** $6t^2 + 3t + at$

28 $96mt^2 - 3mt + 69m^2t$ **29** $6ab^2 + 2ab - 4a^2b$ **30** $5pt^2 + 15pt + 5p^2t$

Factorise the following expressions where possible. List those which cannot factorise.

31 $5m - 6t$ **32** $3m + 2mp$ **33** $t^2 - 5t$ **34** $6pt + 5ab$

35 $8m^2 - 6mp$ **36** $a^2 + c$ **37** $3a^2 - 7ab$ **38** $4ab + 5cd$

39 $7ab - 4b^2c$ **40** $3p^2 - 4t^2$ **41** $6m^2t + 9t^2m$ **42** $5mt + 3pn$

HOMEWORK 23G

Expand the following expressions.

1 $(x + 2)(x + 5)$ **2** $(t + 3)(t + 2)$ **3** $(w + 4)(w + 1)$

4 $(m + 6)(m + 2)$ **5** $(k + 2)(k + 4)$ **6** $(a + 3)(a + 1)$

7 $(x + 3)(x - 1)$ **8** $(t + 6)(t - 4)$ **9** $(w + 2)(w - 3)$

10 $(f + 1)(f - 4)$ **11** $(g + 2)(g - 5)$ **12** $(y + 5)(y - 2)$

13 $(x - 4)(x + 3)$ **14** $(p - 3)(p + 2)$ **15** $(k - 5)(k + 1)$

16 $(y - 3)(y + 6)$ **17** $(a - 2)(a + 4)$ **18** $(t - 4)(t + 5)$

19 $(x - 3)(x - 2)$ **20** $(r - 4)(r - 1)$ **21** $(m - 1)(m - 7)$

22 $(g - 5)(g - 3)$ **23** $(h - 6)(h - 2)$ **24** $(n - 2)(n - 8)$

25 $(4 + x)(3 + x)$ **26** $(5 + t)(4 - t)$ **27** $(2 - b)(6 + b)$

28 $(7 - y)(5 - y)$ **29** $(3 + p)(p - 2)$ **30** $(3 - k)(k - 5)$

31 $(x + 3)^2$ **32** $(t + 4)^2$ **33** $(q - 1)^2$

34 $(k - 5)^2$ **35** $(x + 4)(x - 4)$ **36** $(6 + y)(6 - y)$

HOMEWORK 23H

Example The expression $3x + 2$ has the value 5 when $x = 1$ and 14 when $x = 4$.

1 Find the value of $2x + 3$ when:
a $x = 2$ **b** $x = 5$ **c** $x = 10$.

2 Find the value of $3k - 4$ when:
a $k = 2$ **b** $k = 6$ **c** $k = 12$.

3 Find the value of $4 + t$ when:
a $t = 4$ **b** $t = 20$ **c** $t = \frac{1}{2}$.

4 Evaluate $10 - 2x$ when:
a $x = 3$ **b** $x = 5$ **c** $x = 6$.

5 Evaluate $5y + 10$ when:

a $y = 5$ **b** $y = 10$ **c** $y = 15$.

6 Evaluate $6d - 2$ when:

a $d = 2$ **b** $d = 5$ **c** $d = \frac{1}{2}$.

7 Find the value of $\frac{x + 2}{4}$ when:

a $x = 6$ **b** $x = 10$ **c** $x = 18$.

8 Find the value of $\frac{3x - 1}{2}$ when:

a $x = 1$ **b** $x = 3$ **c** $x = 4$.

9 Evaluate $\frac{20}{p}$ when:

a $p = 2$ **b** $p = 10$ **c** $p = 20$.

10 Find the value of $3(2y + 5)$ when:

a $y = 1$ **b** $y = 3$ **c** $y = 5$.

HOMEWORK 23I

Example The formula for the perimeter of a rectangle is $P = 2a + 2b$. Find P when $a = 5$ and $b = 3$.

$$P = 2 \times 5 + 2 \times 3 = 10 + 6 = 16$$

1 For $A = t + h$, find A when:

a $t = 2$ and $h = 3$ **b** $t = 4$ and $h = 7$ **c** $t = 10$ and $h = 19$.

2 For $P = 2x - 4y$, find P when:

a $x = 5$ and $y = 2$ **b** $x = 6$ and $y = 1$ **c** $x = 8$ and $y = 4$.

3 For $a = 3b + 5c$, find a when:

a $b = 2$ and $c = 3$ **b** $b = 3$ and $c = 5$ **c** $b = 2$ and $c = -2$.

4 For $e = f^2 + g^2$, find e when:

a $f = 2$ and $g = 3$ **b** $f = 3$ and $g = 4$ **c** $f = 5$ and $g = 10$.

5 For $y = \sqrt{x} + n$, find y when:

a $x = 16$ and $n = 5$ **b** $x = 49$ and $n = 3$ **c** $x = 100$ and $n = 50$.

6 The formula to find the distance travelled in miles (d) is given by $d = st$ where s is the average speed in miles per hour and t is the time in hours. Find d when:

a $s = 3$ and $t = 2$ **b** $s = 50$ and $t = 4$ **c** $s = 85$ and $t = 6$.

7 The formula $W = B + RT$ can be used to calculate a person's wage, where W is the total wage, B is the bonus, R is the rate of pay per hour and T is the number of hours worked. Find W when:

a $B = 10$, $R = 6$ and $T = 7$ **b** $B = 40$, $R = 25$ and $T = 40$.

8 The formula for the area of a trapezium is given by $A = \frac{(a + b)h}{2}$.

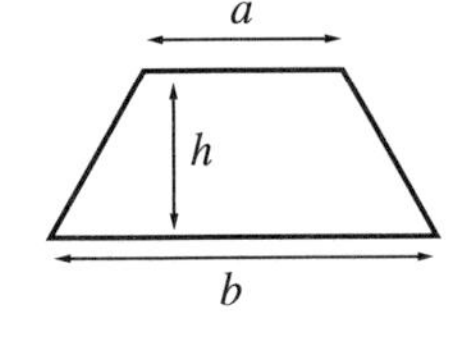

Find the area of a trapezium when:

a $a = 3$, $b = 4$ and $h = 2$

b $a = 5$, $b = 7$ and $h = 10$.

24 Equations and inequalities

HOMEWORK 24A

Solve the following equations.

1 $x + 2 = 8$ **2** $y - 4 = 3$ **3** $s + 7 = 10$ **4** $t - 7 = 4$

5 $3p = 12$ **6** $5q = 15$ **7** $\frac{k}{2} = 4$ **8** $4n = 20$

9 $\frac{a}{3} = 2$ **10** $b + 1 = 2$ **11** $c - 7 = 7$ **12** $\frac{d}{5} = 1$

HOMEWORK 24B

Example Solve $3x - 4 = 11$ using an inverse flow diagram.

The flow diagram for the equation is:

$x \longrightarrow \boxed{\times 3} \longrightarrow \boxed{-4} \longrightarrow 11$

Inverse flow diagram:

$x \longleftarrow \boxed{\div 3} \longleftarrow \boxed{+4} \longleftarrow 11$

Put through the value on the right-hand side:

$5 \longleftarrow \boxed{\div 3} \longleftarrow \boxed{+4} \longleftarrow 11$

The answer is $x = 5$

Checking the answer gives $3 \times 5 - 4 = 11$ which is correct.

Solve each of the following equations using inverse flow diagrams. Do not forget to check that each answer works in the original equation.

1 $2x + 5 = 13$ **2** $3x - 2 = 4$ **3** $2x - 7 = 3$ **4** $3y - 9 = 9$

5 $5a + 1 = 11$ **6** $4x + 5 = 21$ **7** $6y + 6 = 24$ **8** $5x + 4 = 9$

9 $8x - 10 = 30$ **10** $\frac{x}{2} + 1 = 4$ **11** $\frac{a}{2} - 2 = 3$ **12** $\frac{c}{3} + 2 = 8$

13 $\frac{x}{3} - 3 = 1$ **14** $\frac{m}{3} - 1 = 2$ **15** $\frac{z}{5} + 6 = 10$

HOMEWORK 24C

Example Solve the equation $3x - 5 = 16$ by 'doing the same to both sides'.

$3x - 5 = 16$ Add 5 to both sides

$3x - 5 + 5 = 16 + 5$

$3x = 21$ Divide both sides by 3

$\frac{3x}{3} = \frac{21}{3}$

$x = 7$

Solve each of the following equations by 'doing the same to both sides'. Do not forget to check that each answer works in the original equation.

1 $x + 5 = 6$	**2** $y - 3 = 4$	**3** $x + 5 = 3$	**4** $2y + 4 = 12$
5 $3t + 5 = 20$	**6** $2x - 4 = 12$	**7** $6b + 3 = 21$	**8** $4x + 1 = 5$
9 $2m - 3 = 4$	**10** $\frac{x}{2} - 5 = 2$	**11** $\frac{a}{3} + 3 = 6$	**12** $\frac{z}{5} - 1 = 1$

HOMEWORK 24D

Example Solve $4x + 3 = 23$.

Subtract 3 to give $4x = 23 - 3 = 20$
Now divide both sides by 4 to give $x = 20 \div 4 = 5$
The solution is $x = 5$

Solve each of the following equations. Do not forget to check that each answer works in the original equation.

1 $2x + 1 = 7$	**2** $2t + 5 = 13$	**3** $3x + 5 = 17$	**4** $4y + 7 = 27$
5 $2x - 8 = 12$	**6** $5t - 3 = 27$	**7** $\frac{x}{2} + 3 = 6$	**8** $\frac{p}{3} + 2 = 3$
9 $\frac{x}{2} - 3 = 5$	**10** $8 - x = 2$	**11** $13 - 2k = 3$	**12** $6 - 3z = 0$
13 $\frac{x + 2}{3} = 4$	**14** $\frac{y - 4}{5} = 2$	**15** $\frac{z + 4}{8} = 5$	

HOMEWORK 24E

Example Solve $3(2x - 7) = 15$.

First multiply out the bracket to get
$6x - 21 = 15$ Add 21 to both sides
$6x = 36$ Divide both sides by 6
$x = 6$

Solve each of the following equations. Some of the answers may be decimals or negative numbers. Remember to check that each answer works in the original equation. Use your calculator if necessary.

1 $2(x + 1) = 8$	**2** $3(x - 3) = 12$	**3** $3(t + 2) = 9$	**4** $2(x + 5) = 20$
5 $2(2y - 5) = 14$	**6** $2(3x + 4) = 26$	**7** $4(3t - 1) = 20$	**8** $2(t + 5) = 6$
9 $2(x + 4) = 2$	**10** $2(3y - 2) = 5$	**11** $4(3k - 1) = 11$	**12** $5(2x + 3) = 26$

HOMEWORK 24F

Example Solve $5x + 4 = 3x + 10$.

Subtract $3x$ from both sides $2x + 4 = 10$
Subtract 4 from both sides $2x = 6$
Divide both sides by 2 $x = 3$

Solve each of the following equations.

1 $2x + 1 = x + 3$ **2** $3y + 2 = 2y + 6$ **3** $5a - 3 = 4a + 4$

4 $5t + 3 = 3t + 9$ **5** $7p - 5 = 5p + 3$ **6** $6k + 5 = 3k + 20$

7 $6m + 1 = m + 11$ **8** $5s - 1 = 2s - 7$ **9** $4w + 8 = 2w + 8$

10 $5x + 5 = 3x + 10$ **11** $5(t - 2) = 4t - 1$ **12** $4(x + 2) = 2(x + 1)$

13 $5p - 2 = 5 - 2p$ **14** $2(2x + 3) = 3(x - 4)$

15 $5(2y - 1) - 2y = 2(3y - 4) + 10$

HOMEWORK 24G

Set up an equation to represent each situation described below. Then solve the equation. Do not forget to check each answer.

1 A girl is Y years old. Her father is 23 years older than she is. The sum of their ages is 37. How old is the girl?

2 A boy is X years old. His sister is twice as old as he is. The sum of their ages is 24. How old is the boy?

3 The diagram shows a rectangle.
Find x if the perimeter is 24 cm.

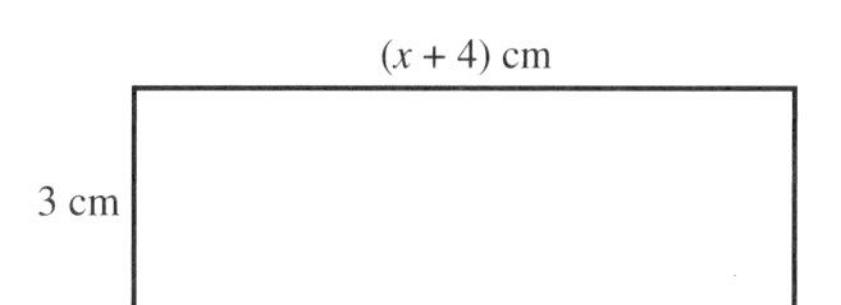

4 Find the length of each side of the pentagon, if it has a perimeter of 32 cm.

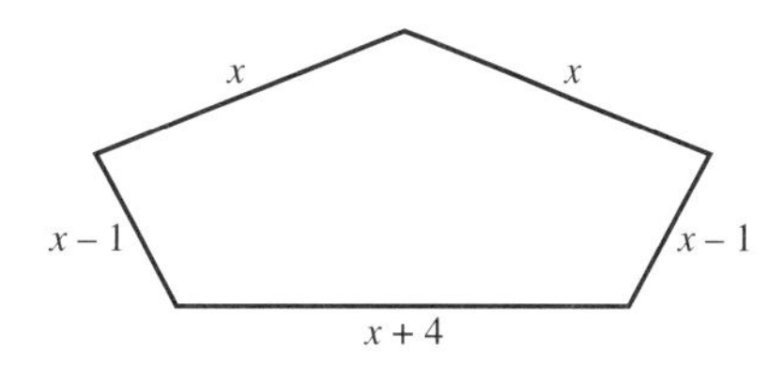

5 On a bookshelf there are $2b$ crime novels, $3b - 2$ science fiction novels and $b + 7$ romance novels. Find how many of each type of book there is, if there are 65 books altogether.

★**6** Maureen thought of a number. She multiplied it by 4 and then added 6 to get an answer of 26. What number did she start with?

★**7** Declan also thought of a number. He took away 4 from the number and then multiplied by 3 to get an answer of 24. What number did he start with?

★**8** Sandeep's money box contains 50p coins, £1 coins and £2 coins.
In the box there are twice as many £1 coins than 50p coins and 4 more £2 coins than 50p coins. There are 44 coins in the box.
a Find how many of each coin there is in the box.
b How much money does Sandeep have in her money box?

HOMEWORK 24H

1 Without using a calculator, find the two consecutive whole numbers between which the solution to each of the following equations lies.
a $x^3 = 10$ **b** $x^3 = 50$ **c** $x^3 = 800$ **d** $x^3 = 300$

2 Show that $x^2 + 2x = 20$ has a solution between $x = 3$ and $x = 4$, and find the solution to 1 decimal place.

3 Find two consecutive whole numbers between which the solution to each of the following equations lies.

a $x^3 + x = 7$ **b** $x^3 + x = 55$ **c** $x^3 + x = 102$ **d** $x^3 + x = 89$

4 Find a solution to each of the following equations to 1 decimal place.

a $x^3 - x = 30$ **b** $x^3 - x = 95$ **c** $x^3 - x = 150$ **d** $x^3 - x = 333$

5 Show that $x^3 + x = 45$ has a solution between $x = 3$ and $x = 4$, and find the solution to 1 decimal place.

6 Show that $x^3 - 2x = 95$ has a solution between $x = 4$ and $x = 5$, and find the solution to 1 decimal place.

HOMEWORK 24I

1 $y = mx + c$ — **i** Make c the subject. **ii** Express x in terms of y, m and c.

2 $v = u - 10t$ — **i** Make u the subject. **ii** Express t in terms of v and u.

3 $T = 2x + 3y$ — **i** Express x in terms of T and y. **ii** Make y the subject.

4 $p = q^2$ — Make q the subject.

5 $p = q^2 - 3$ — Make q the subject.

6 $a = b^2 + c$ — Make b the subject.

★**7** A rocket is fired vertically upwards with an initial velocity of u metres per second. After t seconds the rocket's velocity, v metres per second, is given by the formula $v = u + gt$, where g is a constant.

a Calculate v when $u = 120$, $g = -9.8$ and $t = 6$.

b Rearrange the formula to express t in terms of v, u, and g.

c Calculate t when $u = 100$, $g = -9.8$ and $v = 17.8$.

HOMEWORK 24J

1 Solve the following linear inequalities.

a $x + 3 < 8$ **b** $t - 2 > 6$ **c** $p + 3 \geqslant 11$

d $4x - 5 < 7$ **e** $3y + 4 \leqslant 22$ **f** $2t - 5 > 13$

g $\frac{x+3}{2} < 8$ **h** $\frac{y+4}{3} \leqslant 5$ **i** $\frac{t-2}{5} \geqslant 7$

j $2(x - 3) < 14$ **k** $4(3x + 2) \leqslant 32$ **l** $5(4t - 1) \geqslant 30$

m $3x + 1 \geqslant 2x - 5$ **n** $6t - 5 \leqslant 4t + 3$ **o** $2y - 11 \leqslant y - 5$

p $3x + 2 \geqslant x + 3$ **q** $4w - 5 \leqslant 2w + 2$ **r** $2(5x - 1) \leqslant 2x + 3$

2 Write down the values of x that satisfy each of the following.

a $x - 2 \leqslant 3$, where x is a positive integer.

b $x + 3 < 5$, where x is a positive, even integer.

c $2x - 14 < 38$, where x is a square number.

d $4x - 6 \leqslant 15$, where x is a positive, odd number.

e $2x + 3 < 25$, where x is a positive, prime number.

HOMEWORK 24K

1 Write down the inequality that is represented by each diagram below.

a

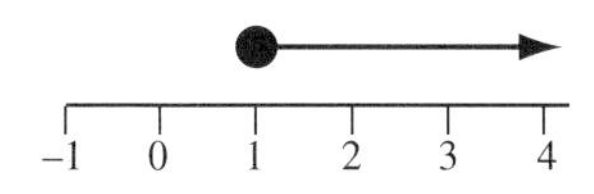

b

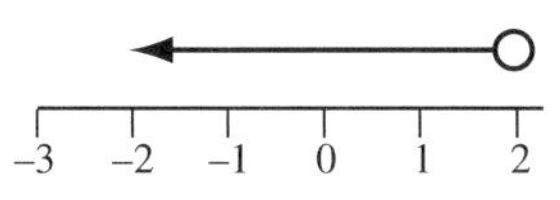

c

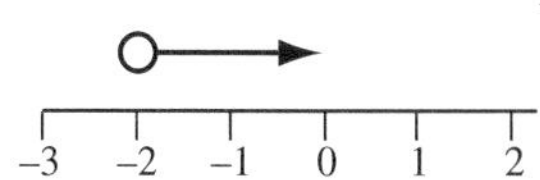

d

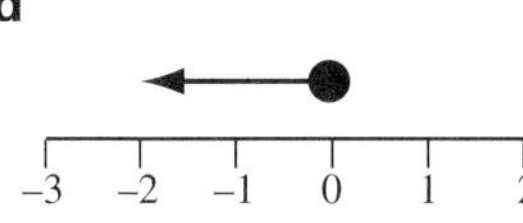

e

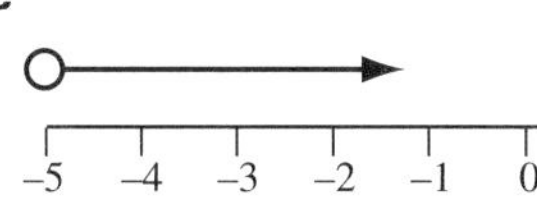

f

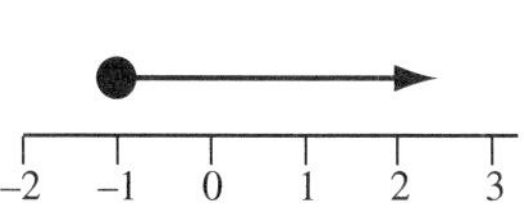

2 Draw diagrams to illustrate the following.

a $x \leqslant 2$ **b** $x > -3$ **c** $x \geqslant 0$ **d** $x < 4$

e $x \geqslant -3$ **f** $1 < x \leqslant 4$ **g** $-2 \leqslant x \leqslant 4$ **h** $-2 < x < 3$

3 Solve the following inequalities and illustrate their solutions on number lines.

a $x + 5 \geqslant 9$ **b** $x + 4 < 2$ **c** $x - 2 \leqslant 3$ **d** $x - 5 > -2$

e $4x + 3 \leqslant 9$ **f** $5x - 4 \geqslant 16$ **g** $2x - 1 > 13$ **h** $3x + 6 < 3$

i $3(2x + 1) < 15$ **j** $\frac{x + 1}{2} \leqslant 2$ **k** $\frac{x - 3}{3} > 7$ **l** $\frac{x + 6}{4} \geqslant 1$

25 Graphs

HOMEWORK 25A

1 A hire firm hired out large scanners. They used the following graph to approximate what the charges would be.

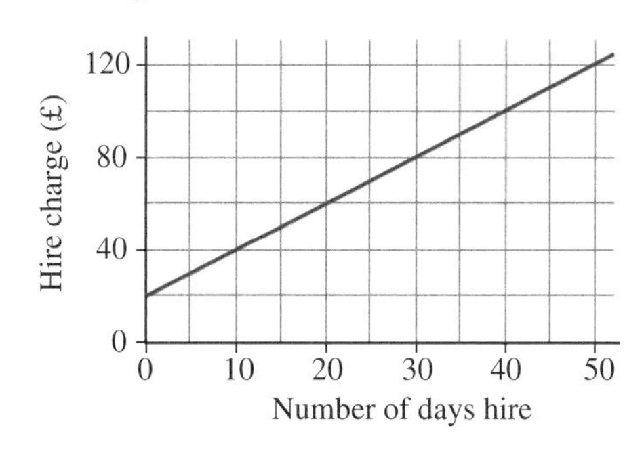

a Use the graph to find the approximate charge for hiring a scanner for:
i 20 days **ii** 30 days **iii** 50 days.

b Use the graph to find out how many days hire you would get for a cost of:
i £120 **ii** £100 **iii** £70.

2 A conference centre used the following chart for the approximate cost of a conference based on the number of people attending it.

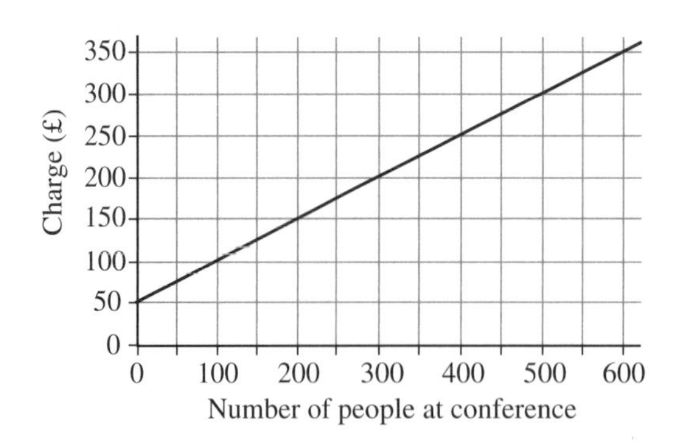

a Use the graph to find the approximate charge for:
i 500 people **ii** 300 people **iii** 250 people.

b Use the graph to estimate how many people can attend a conference at the centre for a cost of:
i £250 **ii** £150 **iii** £125.

3 Jayne lost her fuel bill, but while talking to her friends, she found out that:
Kris who had used 750 units was charged £69
Nic who had used 250 units was charged £33
Shami who had used 500 units was charged £51.

a Plot the given information and draw a straight-line graph. Use a scale from 0 to 800 on the horizontal units axis, and from £0 to £70 on the vertical cost axis.

b Use your graph to find what Jayne will be charged for 420 units.

HOMEWORK 25B

1 Joe was travelling in his car to meet his girlfriend. He set off from home at 9.00 pm, and stopped on the way for a break. This distance–time graph illustrates his journey.

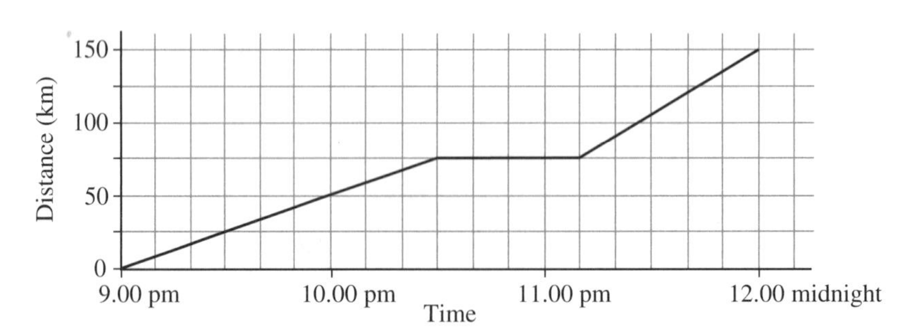

a At what time did he:
i stop for his break **ii** set off after his break **iii** get to his meeting place?

b At what average speed was he travelling:
i over the first hour **ii** over the last hour **iii** for the whole of his journey?

2 Jean set off in a taxi from Hellaby. The taxi then went on to pick up Jeans's parents. It then travelled further, dropping them all off at a shopping centre. The taxi went on a further 10 km to pick up another party and took them back to Hellaby. This distance–time graph illustrates the journey.

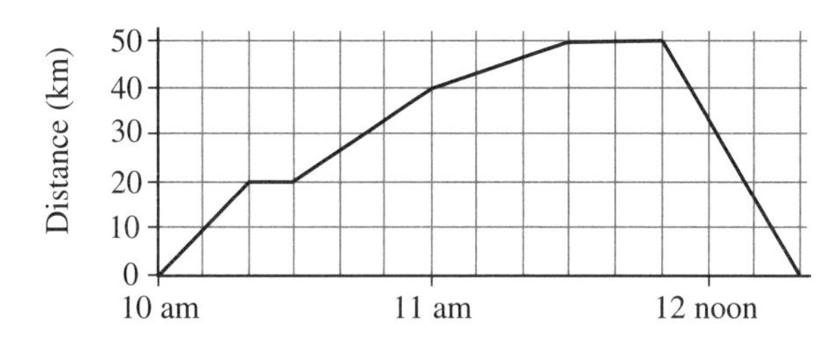

- **a** How far from Hellaby did Jean's parents live?
- **b** How far from Hellaby is the shopping centre?
- **c** What was the average speed of the taxi while only Jean was in the taxi?
- **d** What was the average speed of the taxi back to Hellaby?

3 Grandad took his grandchildren out for a trip. He set off at 1.00 pm and travelled, for half an hour, away from Norwich at an average speed of 60 km/h. They stopped to look at the sea and have an ice cream. At two o'clock, they set off again, travelling for a quarter of an hour at a speed of 80 km/h. Then they stopped to play on the sand for half an hour. Grandad then drove the grandchildren back home at an average speed of 50 km/h. Draw a travel graph to illustrate this story. Use a horizontal axis to represent time from 1 pm to 5 pm, and a vertical scale from 0 km to 60 km.

HOMEWORK 25C

1 Draw the graph of $y = x + 1$

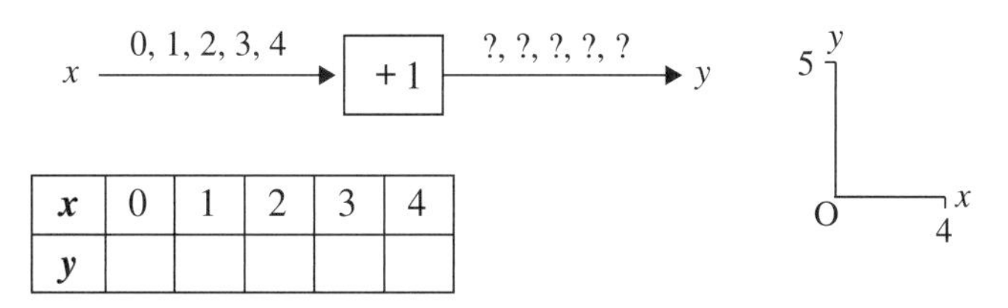

x	0	1	2	3	4
y					

2 Draw the graph of $y = 2x + 1$

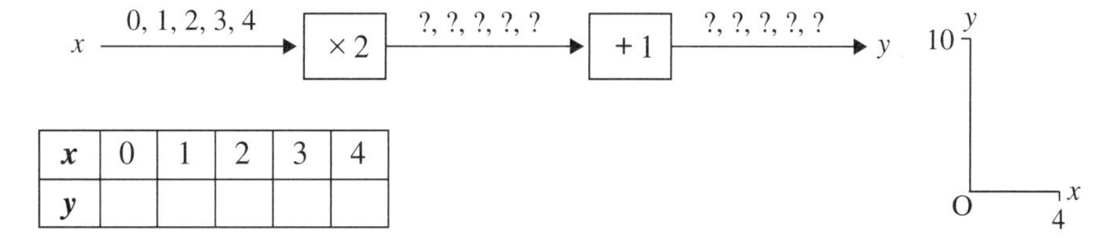

x	0	1	2	3	4
y					

3 Draw the graph of $y = 3x + 1$

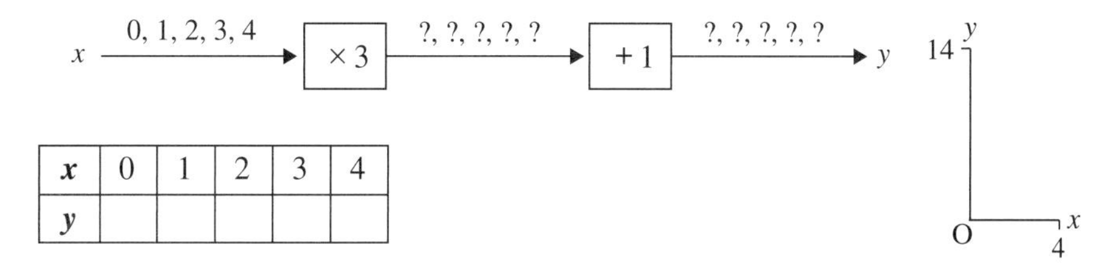

x	0	1	2	3	4
y					

4 Draw the graph of $y = x - 1$

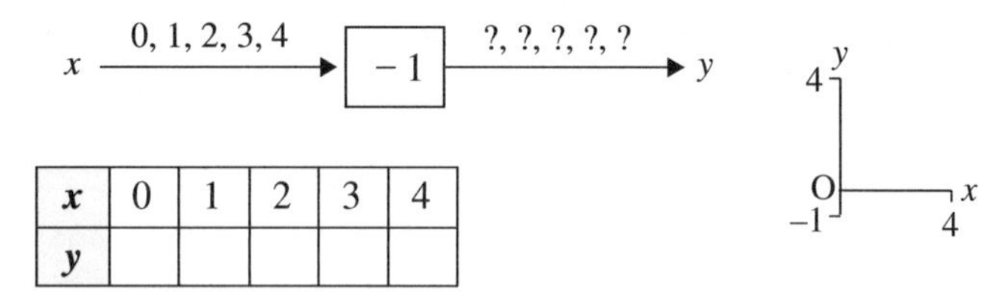

x	0	1	2	3	4
y					

★**5** **a** Draw the graphs of $y = x - 2$ and $y = 2x - 1$ on the same grid.
b Where do the graphs cross?

★**6** **a** Draw the graphs of $y = 2x$ and $y = x + 2$ on the same grid.
b Where do the graphs cross?

HOMEWORK 25D

Draw the graph for each of the equations given.

Follow these hints.

- Use the highest and smallest values of x given as your range.
- When the first part of the function is a division, pick x-values that divide exactly to avoid fractions.
- Always label your graphs. This is particularly important when you are drawing two graphs on the same set of axes.
- Create a table of values. You will often have to complete these in your examinations.

1 Draw the graph of $y = 2x + 3$ for x-values from 0 to 5 $(0 \leqslant x \leqslant 5)$

2 Draw the graph of $y = 3x - 1$ $(0 \leqslant x \leqslant 5)$

3 Draw the graph of $y = \frac{x}{2} - 2$ $(0 \leqslant x \leqslant 12)$

4 Draw the graph of $y = 2x + 1$ $(-2 \leqslant x \leqslant 2)$

5 Draw the graph of $y = \frac{x}{2} + 5$ $(-6 \leqslant x \leqslant 6)$

6 **a** On the same set of axes, draw the graphs of
$y = 3x - 1$ and $y = 2x + 3$ $(0 \leqslant x \leqslant 5)$
b Where do the two graphs cross?

7 **a** On the same axes, draw the graphs of
$y = 4x - 3$ and $y = 3x + 2$ $(0 \leqslant x \leqslant 6)$
b Where do the two graphs cross?

8 **a** On the same axes, draw the graphs of
$y = \frac{x}{2} + 1$ and $y = \frac{x}{3} + 2$ $(0 \leqslant x \leqslant 12)$
b Where do the two graphs cross?

9 **a** On the same axes, draw the graphs of
$y = 2x + 3$ and $y = 2x - 1$ $(0 \leqslant x \leqslant 4)$
b Do the graphs cross? If not, why not?

10 **a** Copy and complete the table to draw the graph of
$x + y = 6$ $(0 \leqslant x \leqslant 6)$
b Now draw the graph of $x + y = 3$ $(0 \leqslant x \leqslant 6)$

x	0	1	2	3	4	5	6
y							

26 Pattern

HOMEWORK 26A

Look for the pattern and then write the next two lines. Check your answers with a calculator afterwards.

1 $7 \times 11 \times 13 \times 2 = 2002$
$7 \times 11 \times 13 \times 3 = 3003$
$7 \times 11 \times 13 \times 4 = 4004$
$7 \times 11 \times 13 \times 5 = 5005$

2 $3 \times 7 \times 13 \times 37 \times 2 = 20202$
$3 \times 7 \times 13 \times 37 \times 3 = 30303$
$3 \times 7 \times 13 \times 37 \times 4 = 40404$
$3 \times 7 \times 13 \times 37 \times 5 = 50505$

3 $3 \times 5 = 4^2 - 1 = 15$
$4 \times 6 = 5^2 - 1 = 24$
$5 \times 7 = 6^2 - 1 = 35$
$6 \times 8 = 7^2 - 1 = 48$

4 $3 \times 7 = 5^2 - 4 = 21$
$4 \times 8 = 6^2 - 4 = 32$
$5 \times 9 = 7^2 - 4 = 45$
$6 \times 10 = 8^2 - 4 = 60$

From your observations on the number patterns above, answer Questions **5** to **9** without using a calculator. Check with a calculator once you have attempted them.

5 $7 \times 11 \times 13 \times 9 =$

6 $3 \times 7 \times 13 \times 37 \times 8 =$

7 $7 \times 11 \times 13 \times 15 =$

8 $3 \times 7 \times 13 \times 37 \times 15 =$

9 $3 \times 7 \times 13 \times 37 \times 99 =$

HOMEWORK 26B

1 Look at the following number sequences. Write down the next three terms in each and explain how each sequence is found.

a 4, 6, 8, 10, …
b 3, 6, 9, 12, …
c 2, 4, 8, 16, …
d 5, 12, 19, 26, …
e 3, 30, 300, 3000, …
f 1, 4, 9, 16, …

2 Look carefully at each number sequence below. Find the next two numbers in the sequence and try to explain the pattern.

a 1, 2, 3, 5, 8, 13, 21, …
b 2, 3, 5, 8, 12, 17, …

3 Look at the sequences below. Find the rule for each sequence and write down its next three terms.

a 7, 14, 28, 56, …
b 3, 10, 17, 24, 31, …
c 1, 3, 7, 15, 31, …
d 40, 39, 37, 34, …
e 3, 6, 11, 18, 27, …
f 4, 5, 7, 10, 14, 19, …
g 4, 6, 7, 9, 10, 12, …
h 5, 8, 11, 14, 17, …
i 5, 7, 10, 14, 19, 25, …
j 10, 9, 7, 4, …
k 200, 40, 8, 1.6, …
l 3, 1.5, 0.75, 0.375, …

HOMEWORK 26C

1 Use each of the following rules to write down the first five terms of a sequence.

a $3n + 1$ for $n = 1, 2, 3, 4, 5$
b $2n - 1$ for $n = 1, 2, 3, 4, 5$
c $4n + 2$ for $n = 1, 2, 3, 4, 5$
d $2n^2$ for $n = 1, 2, 3, 4, 5$
e $n^2 - 1$ for $n = 1, 2, 3, 4, 5$

2 Write down the first five terms of the sequence which has its nth term as:

a $n + 2$ **b** $4n - 1$ **c** $4n - 3$ **d** $n^2 + 1$ **e** $2n^2 + 1$.

HOMEWORK 26D

1 Find the *n*th term in each of these linear sequences.

a 5, 7, 9, 11, 13 … **b** 3, 7, 11, 15, 19, … **c** 6, 11, 16, 21, 26, …

d 3, 9, 15, 21, 27, … **e** 4, 7, 10, 13, 16, … **f** 3, 10, 17, 24, 31, …

2 Find the 50th term in each of these linear sequences.

a 3, 5, 7, 9, 11, … **b** 5, 9, 13, 17, 21, … **c** 8, 13, 18, 23, 28, …

d 2, 8, 14, 20, 26, … **e** 5, 8, 11, 14, 17, … **f** 2, 9, 16, 23, 30, …

3 For each sequence **a** to **f**, find:

i the *n*th term **ii** the 100th term **iii** the term closest to 100.

a 4, 7, 10, 13, 16, … **b** 7, 9, 11, 13, 15, … **c** 3, 8, 13, 18, 23, …

d 1, 5, 9, 13, 17, … **e** 2, 10, 18, 26, … **f** 5, 6, 7, 8, 9, …

4 p is an odd number and q is an even number. State whether the following are odd or even.

a $p + 5$ **b** $q - 3$ **c** $2p$ **d** q^2

e pq **f** $2(p + q)$ **g** $p^2 + q$ **h** $q(p + q)$

HOMEWORK 26E

1 A conference centre had tables each of which could sit 3 people. When put together, the tables could seat people as shown.

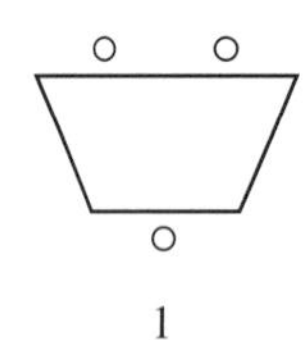

1 2

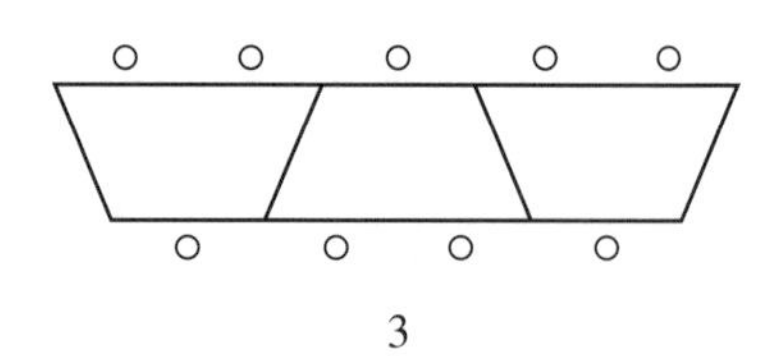

3

a How many people could be seated at 4 tables?

b How many people could be seated at *n* tables put together in this way?

c A conference had 50 people who wished to use the tables in this way. How many tables would they need?

2 A pattern of shapes is built up from matchsticks as shown.

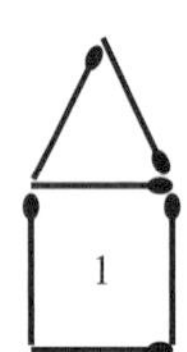

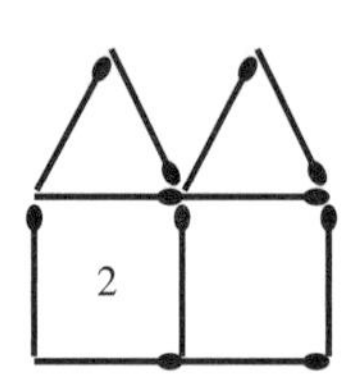

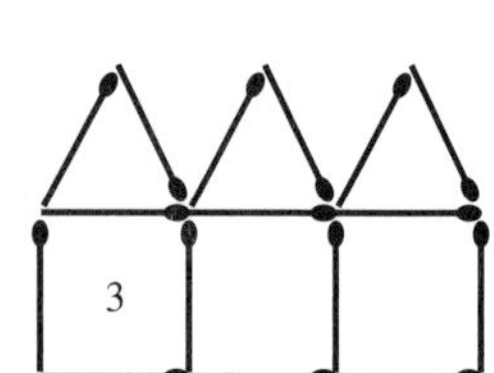

a Draw the 4th diagram.

b How many matchsticks are in the *n*th diagram?

c How many matchsticks are in the 25th diagram?

d With 200 matchsticks, which is the biggest diagram that could be made?

3 A pattern of hexagons is built up from matchsticks.

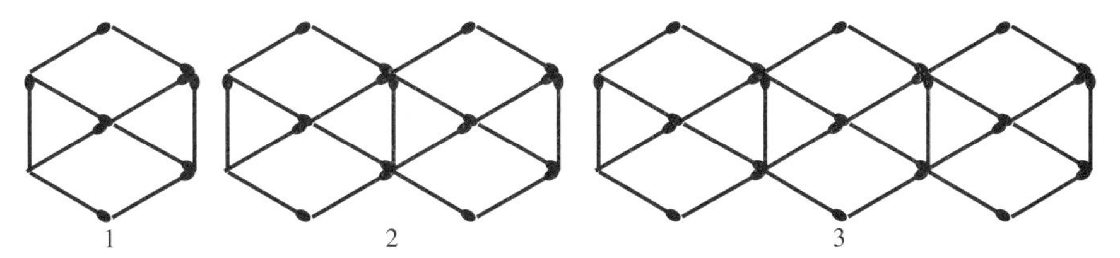

a Draw the 4th set of hexagons in this pattern.

b How many matchsticks are needed for the nth set of hexagons?

c How many matchsticks are needed to make the 60th set of hexagons?

d If there are only 100 matchsticks, which is the largest set of hexagons that could be made?